Erwin Dee Kord (Ed.)

Ostkreuz

Erwin Dee Kord (Ed.)

Ostkreuz

Berlin S-Bahn, East Berlin, Friedrichshain

Solv

Imprint

Publisher:
Solv is a trademark of
International Book Market Service Ltd., 17 Rue Meldrum, Beau Bassin, 1713-01 Mauritius
Email: info@bookmarketservice.com
Website: www.bookmarketservice.com

Published in 2011

Printed in: U.S.A., U.K., Germany. This book was not produced in Mauritius.

ISBN: 978-613-7-97932-7

Contents

Articles

Ostkreuz	1
Berlin	4
Berlin S-Bahn	33
Commuter rail	40
East Berlin	47
Friedrichshain	52
Friedrichshain-Kreuzberg	57
Rummelsburg	59
Berlin Westkreuz station	62
Berlin Stadtbahn	64
Berlin Ringbahn	72
Berlin Wall	76
West Berlin	101
German reunification	120
East Germany	129
Treptower Park	153
Berlin Warschauer Straße station	154

References

Article Sources and Contributors	156
Image Sources, Licenses and Contributors	160

Ostkreuz

Berlin Ostkreuz (German for "East Cross") is a station on the Berlin S-Bahn suburban railway, and one of the busiest in Germany. The station is in the former East Berlin district of Friedrichshain, now part of the borough of Friedrichshain-Kreuzberg.

Ostkreuz is almost unchanged since the 1920s. This view is looking eastward along Platform D, with Platform F visible above the train.

Overview

The station was opened with the Stadtbahn in 1882 under the name *Stralau-Rummelsburg*. It was given its present name in 1933. Along with Westkreuz it has long served as the interchange point between the Ringbahn that circumnavigates central Berlin and the Stadtbahn that passes east-west through it. During the division of Berlin, when links between East and West Berlin public transit were severed, Ostkreuz served as the main hub of the East Berlin S-Bahn network.

A view of Ostkreuz from above. This view looks north along the Ringbahn, with the Stadtbahn visible on the left (west). The single line closest the camera connects the two, served by Platform A just off the left edge of the image. Platform F on the Ringbahn is centered, with the white roof. The large dark-colored tower was built to store water for steam trains previously running on these lines.

Ostkreuz is notable for its appearance: its buildings and track layout have remained essentially unchanged since the 1920s, and its dilapidated state has earned it the nickname *Rostkreuz* ("Rust Cross"). A modernization of the station was proposed as early as 1937, and again during East German rule. Since German reunification, plans have been made to modernize Ostkreuz from its present, partly dilapidated state, and turn it into a modern station with platforms for regional rail trains and stops for buses and the tramway, both of which pass nearby but presently do not stop directly at the station. However, so far none of these have been implemented, due to the high cost involved, as well as the need to minimize disruption to the station's 100,000 daily passengers, and the fact that the station is under historic preservation and parts of it, such as the brick Ringbahn viaduct and the western pedestrian bridge, will need to be retained. The current planning calls for a start date of mid-2007 for the renovation, although preliminary work began early in the year with the removal of several trees and some temporary structures that had been located at the site.

Operations

The station's six platforms are:

- *Platform A*, is a triangular platform on the eastern end of the station lying between the curved lines connecting the Stadtbahn with the Ringbahn. The northern and southern faces of the platform handled traffic crossing from the westbound Stadtbahn onto the northern or southern Ringbahn (respectively). The northern connection was disused and eventually removed in 2006, leaving only the southern face of A in use, serving trains moving north from Treptower Park and then turning west to Warschauer Straße.
- *Platforms B and C* were the outside platforms opposite the northern and southern faces of A (respectively), serving traffic from the Ringbahn heading eastward onto the Stadtbahn. They became dilapidated and were closed during East German rule, leaving service for westbound traffic only.
- *Platform D and E*, the Stadtbahn platforms (each with two tracks), located at ground level, one for each of the two sets of lines that diverge east of the station. D is north of E and serves the lines that eastboand travel towards Strausberg, Wartenberg, and Ahrensfelde, and the same lines traveling westbound. E is for lines that eastbound travel towards Erkner, and the same lines traveling westbound.
- *Platform F* is the tracks of the Ringbahn, located above.

Due to the station's archaic layout, it (along with nearby Warschauer Straße station, currently being rebuilt) retains a platform-allocation system in which trains stop at its east-west Stadtbahn platforms (D and E) according to their line number, not direction. For example, the next train to Berlin Ostbahnhof could leave from either D, E, or the tangential platform A, and moving from one of these platforms to another requires navigating steps and a pedestrian walkway. It is a common sight to see passengers waiting on the walkway until they see where the next westbound train will arrive.

The renovation plans call for the tangential platforms to be eliminated altogether, so that trains between Treptower Park and Warschauer Straße will not stop in either direction. The northern tangential line will not be restored.

The station code is *BOK*, with the subcodes *BOK F* (Ringbahn platform), *BOK D* (Platform D, Stadtbahn), *BOK E* (Platform E, toward Erkner).

Sources

- Andreas Butter, Hans-Joachim Kirsche und Erich Preuß: *Berlin Ostkreuz – Die Drehscheibe des S-Bahn-Verkehrs*, Geramond Verlag, München, 2000, ISBN 3-932785-24-X

External links

- www.stadtschnellbahn-berlin.de [1] - Information about Ostkreuz and reconstruction plans
- Ostkreuz rebuilding blog (in German) [2]
- *Ostkreuz guide* [3] - Photos of Ostkreuz
- *Umgebungsplan* [4] - S-Bahn Berlin neighborhood map

Preceding station	Berlin S-Bahn	Following station
Warschauer Straße *toward Spandau*	S3	Rummelsburg *toward Erkner*
Frankfurter Allee *One-way operation*	S41	Treptower Park *toward Ringbahn (clockwise)*
Treptower Park *One-way operation*	S42	Frankfurter Allee *toward Ringbahn (counter-clockwise)*
Warschauer Straße *toward Westkreuz*	S5	Nöldnerplatz *toward Strausberg Nord*
Warschauer Straße *toward Potsdam Hbf*	S7	Nöldnerplatz *toward Ahrensfelde*
Warschauer Straße *toward Spandau*	S75	Nöldnerplatz *toward Wartenberg*
Frankfurter Allee *toward Hohen Neuendorf*	S8	Treptower Park *toward Zeuthen*
Frankfurter Allee *toward Waidmannslust*	S85	Treptower Park *toward Grünau*
	(move not since winter 2009)	
Frankfurter Allee *toward Blankenburg*	S9	Treptower Park *toward Flughafen Schönefeld*

References

[1] http://www.stadtschnellbahn-berlin.de/index.php?http://www.stadtschnellbahn-berlin.de/bahnhof/356.html
[2] http://www.ostkreuzblog.de
[3] http://ostxgui.de/
[4] http://www.s-bahn-berlin.de/fahrplanundnetz/pdf_umgebungsplaene/SOstkreuz.pdf

Berlin

Berlin

— State of Germany —

Left to right: Berliner Fernsehturm and Skyline, Siegessäule, Kaiser-Wilhelm-Gedächtniskirche, View over Spree river, and Brandenburger Tor

Flag

Coat of arms

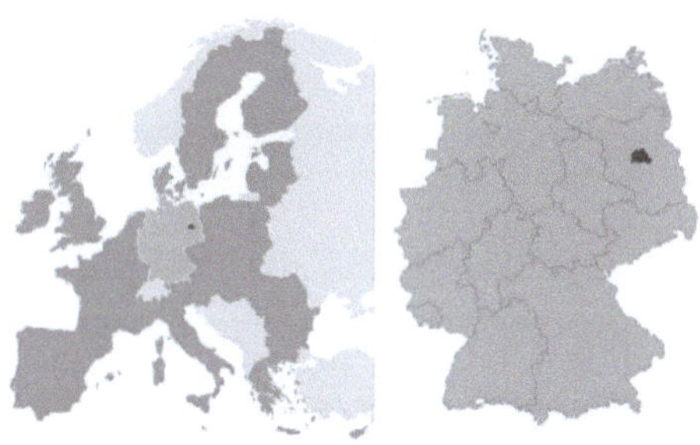

Location within European Union and Germany

Coordinates: 52°30′2″N 13°23′56″E

Country	Germany
Government	
• Governing Mayor	Klaus Wowereit (SPD)
• Governing parties	SPD / CDU
• Votes in Bundesrat	4 (of 69)

Area	
• City	891.85 km^2 (344.3 sq mi)
Elevation	34 m (112 ft)
Population (31 May 2011)[1]	
• City	3479740
• Density	3901.7/km^2 (10105.4/sq mi)
• Metro	4,429,847
Time zone	CET (UTC+1)
• Summer (DST)	CEST (UTC+2)
Postal code(s)	10001–14199
Area code(s)	030
ISO 3166 code	DE-BE
Vehicle registration	B (for earlier signs see note)[2]
GDP/ Nominal	€ 94.7 billion (2010) [3]
NUTS Region	DE3
Website	berlin.de [4]

Berlin (English pronunciation: /bɜrˈlɪn/; German pronunciation: [bɛɐ̯ˈliːn] (listen)) is the capital city of Germany and is one of the 16 states of Germany. With a population of 3.45 million people,[1] Berlin is Germany's largest city. It is the second most populous city proper and the seventh most populous urban area in the European Union.[5] Located in northeastern Germany, it is the center of the Berlin-Brandenburg Metropolitan Region, which has 4.4 million residents from over 190 nations.[6] Located in the European Plains, Berlin is influenced by a temperate seasonal climate. Around one third of the city's area is composed of forests, parks, gardens, rivers and lakes.[7]

First documented in the 13th century, Berlin was the capital of the Kingdom of Prussia (1701–1918), the German Empire (1871–1918), the Weimar Republic (1919–1933) and the Third Reich (1933–1945).[8] Berlin in the 1920s was the third largest municipality in the world.[9] After World War II, the city became divided into East Berlin—the capital of East Germany—and West Berlin, a West German exclave surrounded by the Berlin Wall (1961–1989).[10] Following German reunification in 1990, the city regained its status as the capital of Germany, hosting 147 foreign embassies.[11] [12]

Berlin is a world city of culture, politics, media, and science.[13] [14] [15] Its economy is primarily based on the service sector, encompassing a diverse range of creative industries, media corporations, and convention venues. Berlin also serves as a continental hub for air and rail transport,[16] [17] and is a popular tourist destination.[18] Significant industries include IT, pharmaceuticals, biomedical engineering, biotechnology, electronics, traffic engineering, and renewable energy.

Berlin is home to renowned universities, research institutes, orchestras, museums, and celebrities, as well as host of many sporting events.[19] Its urban settings and historical legacy have made it a popular location for international film productions.[20] The city is well known for its festivals, diverse architecture, nightlife, contemporary arts, public transportation networks and a high quality of living.[21]

History

The origin of the name *Berlin* is unknown, but it may have its roots in the language of West Slavic inhabitants of the area of today's Berlin, and be related to the Old Polabian stem *berl-/birl-* "swamp".[22] Folk etymology connects it to the German *Bär*, a bear, and a bear appears in the coat of arms of the city.

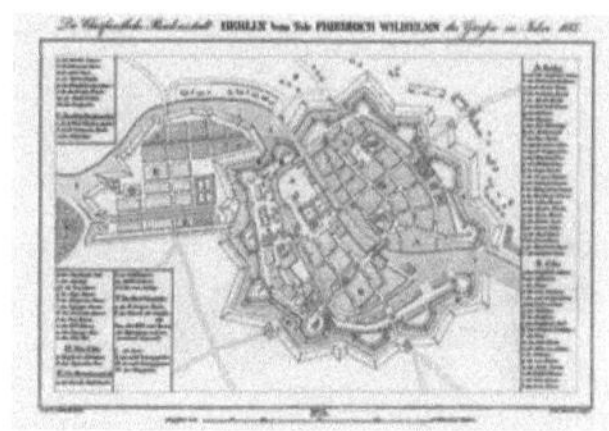

Map of Berlin in 1688

The earliest evidence of settlements in the area of today's Berlin is a wooden beam dated from approximately 1192.[23] The first written records of towns in the area of present-day Berlin date from the late 12th century. Spandau is first mentioned in 1197 and Köpenick in 1209, although these areas did not join Berlin until 1920.[24] The central part of Berlin can be traced back to two towns. Cölln on the Fischerinsel is first mentioned in a 1237 document, and Berlin, across the Spree in what is now called the Nikolaiviertel, is referenced in a document from 1244.[23] The former is considered to be the founding date of the city. The two towns over time formed close economic and social ties and eventually merged in 1307 and came to be known as *Berlin*.

In 1435, Frederick I became the elector of the Margraviate of Brandenburg, which he ruled until 1440.[25] His successor, Frederick II Irontooth, established Berlin as capital of the margraviate, and subsequent members of the Hohenzollern family ruled until 1918 in Berlin, first as electors of Brandenburg, then as kings of Prussia, and eventually as German emperors. In 1448, citizens rebelled in the "Berlin Indignation" against the construction of a new royal palace by Frederick II Irontooth. This protest was not successful, however, and the citizenry lost many of its political and economic privileges. In 1451 Berlin became the royal residence of the Brandenburg electors, and Berlin had to give up its status as a free Hanseatic city. In 1539, the electors and the city officially became Lutheran.[26]

17th to 19th centuries

Frederick the Great (1712–1786) was one of Europe's enlightened monarchs.

The Thirty Years' War between 1618 and 1648 devastated Berlin. One third of its houses were damaged or destroyed, and the city lost half of its population.[27] Frederick William, known as the "Great Elector", who had succeeded his father George William as ruler in 1640, initiated a policy of promoting immigration and religious tolerance. With the Edict of Potsdam in 1685, Frederick William offered asylum to the French Huguenots. More than 15,000 Huguenots went to Brandenburg, of whom 6,000 settled in Berlin. By 1700, approximately 20 percent of Berlin's residents were French, and their cultural influence on the city was immense. Many other immigrants came from Bohemia, Poland, and Salzburg.

Berlin became the capital of the German Empire in 1871 and expanded rapidly in the following years. (Unter den Linden in 1900)

With the coronation of Frederick I in 1701 as king (in Königsberg), Berlin became the new capital of the Kingdom of Prussia (instead of Königsberg); this was a successful attempt to centralize the capital in the very outspread Prussian Kingdom, and it was the first time the city began to grow. In 1740, Frederick II, known as Frederick the Great (1740–1786), came to power. Under the rule of Frederick II Berlin became a center of the Enlightenment. Following France's victory in the War of the Fourth Coalition, Napoleon Bonaparte marched into Berlin in 1806, but granted self-government to the city. In 1815 the city became part of the new Province of Brandenburg.

The Industrial Revolution transformed Berlin during the 19th century; the city's economy and population expanded dramatically, and it became the main rail hub and economic center of Germany. Additional suburbs soon developed and increased the area and population of Berlin. In 1861, outlying suburbs including Wedding, Moabit, and several others were incorporated into Berlin. In 1871, Berlin became capital of the newly founded German Empire. On 1 April 1881 it became a city district separate from Brandenburg.

20th century

At the end of World War I in 1918, a republic was proclaimed in Berlin. In 1920, the Greater Berlin Act incorporated dozens of suburban cities, villages, and estates around Berlin into an expanded city. This new area encompassed Spandau and Charlottenburg in the west, as well as several other areas that are now major municipalities. After this expansion, Berlin had a population of around four million. During the Weimar era, Berlin became internationally renowned as a center of cultural transformation, at the heart of the Roaring Twenties.

Berlin in ruins after World War II (Potsdamer Platz, 1945).

On 30 January 1933, Adolf Hitler and the Nazi Party came to power through the Machtergreifung. Nazi rule destroyed Berlin's Jewish community, which had numbered 170,000 before 1933. After Kristallnacht in 1938, thousands of the city's Jews were imprisoned in the nearby Sachsenhausen concentration camp or, in early 1943, were shipped to death camps, such as Auschwitz. During World War II, large parts of Berlin were destroyed in the 1943–45 air raids and during the Battle of Berlin. Among the hundreds of thousands who died during the Battle for Berlin, an estimated 125,000 were civilians.[28] After the end of the war in Europe in 1945, Berlin received large numbers of refugees from the Eastern provinces. The victorious powers divided the city into four sectors, analogous to the occupation zones into which Germany was divided. The sectors of the Western Allies (the United States, the United Kingdom and France) formed West Berlin, while the Soviet sector formed East Berlin.[29]

The Berlin Wall in 1986, painted on the western side. People crossing the so-called "death strip" on the eastern side were at risk of being shot.

All four Allies shared administrative responsibilities for Berlin. However, in 1948, when the Western Allies extended the currency reform in the Western zones of Germany to the three western sectors of Berlin, the Soviet Union imposed a blockade on the access routes to and from West Berlin, which lay entirely inside Soviet-controlled territory. The Berlin airlift, conducted by the three western Allies, overcame this blockade by supplying food and other supplies to the city from 24 June 1948 to 11 May 1949.[30] In 1949, the Federal Republic of Germany was founded in West Germany and eventually included all of the American, British, and French zones, excluding those three countries' zones in Berlin, while the Marxist-Leninist German Democratic Republic was proclaimed in East Germany. West Berlin officially remained an occupied city, but as a *corpus separatum* it politically was very closely aligned with Federal Republic of Germany despite Berlin's geographic location within East Germany. West Berlin issued its own postage stamps, which were often the same as West German postage stamps but with the additional word "Berlin" added. Airline service to West Berlin was granted only to American, British, and French airlines.

The fall of the Berlin Wall in 1989.

The founding of the two German states increased Cold War tensions. West Berlin was surrounded by East German territory and East Germany proclaimed East Berlin (described as "Berlin") as its capital, a move that was not recognized by the western powers. Although only half the size and population of West Berlin, East Berlin included most of the historic center of the city. The West German government, meanwhile, established itself provisionally in Bonn.[31]

As a result of the political and economical tensions brought on by the Cold War, on 13 August 1961, East Germany began building of the Berlin Wall between East and West Berlin and similar barriers around West Berlin, and events escalated to a tank standoff at Checkpoint Charlie on 27 October 1961. West Berlin was now de facto a part of West Germany with a unique legal status, while East Berlin was de facto a part of East Germany.

Berlin Mitte in the 21st century. Some landmarks from top left to bottom right: Hauptbahnhof (main station), Charité hospital, Berliner Dom, City hall, TV tower at Alexanderplatz, Spree river, East Side Gallery, O2 World.

Berlin was completely divided. Although it was possible for Westerners to pass from one to the other only through strictly controlled checkpoints, for most Easterners, travel to West Berlin or West Germany was no longer possible. In 1971, a Four-Power agreement guaranteed access to and from West Berlin by car or train through East Germany and ended the potential for harassment or closure of the routes.[32]

In 1989, with the end of the Cold War and pressure from the East German population the Berlin Wall fell on 9 November 1989, and was subsequently mostly demolished, with little of its physical structure remaining today; the East Side Gallery in Friedrichshain near the *Oberbaumbrücke* over the Spree preserves a portion of the Wall.

On 3 October 1990, the two parts of Germany were reunified as the Federal Republic of Germany, and Berlin again became the official German capital. In June 1991, the German Parliament, the Bundestag, voted to move the seat of the (West) German capital back from Bonn to Berlin, which was completed in 1999.

Geography

Berlin is located in eastern Germany, approximately 60 km (37 mi) west of the Polish border, in an area of low-lying marshy woodlands with a mainly flat topography, for it is part of the vast Northern European Plain which stretches all the way from northern France to western Russia. The Berlin–Warsaw *Urstromtal* (ice age melt water flow), between the low Barnim plateau to the north and the Teltow plateau to the south, was formed by water flowing from melting ice sheets at the end of the last ice age. The Spree follows this valley now. In Spandau, Berlin's westernmost borough, the Spree meets the river Havel, which flows from north to south through western Berlin. The course of the Havel is more like a chain of lakes, the largest being the Tegeler See and Großer Wannsee. A series of lakes also feeds into the upper Spree, which flows through the Großer Müggelsee in eastern Berlin.[33]

Natural and built environment.

View over central Berlin. Unter den Linden in foreground and skyscrapers of Potsdamer Platz up to the right.

Substantial parts of present-day Berlin extend onto the low plateaus on both sides of the Spree Valley. Large parts of the boroughs Reinickendorf and Pankow lie on the Barnim plateau, while most of the boroughs Charlottenburg-Wilmersdorf, Steglitz-Zehlendorf, Tempelhof-Schöneberg, and Neukölln lie on the Teltow plateau.

The borough of Spandau lies partly within the Berlin *Urstromtal* and partly on the Nauen Plain, which stretches to the west of Berlin. The highest elevations in Berlin are the Teufelsberg and the Müggelberge. Both hills have an elevation of about 115 metres (377 ft). The Teufelsberg is in fact an artificial pile of rubble from the ruins of World War II.

Climate

Berlin has a humid continental climate according to the Köppen climate classification system.

Summers are warm and sometimes humid with average high temperatures of 22–25 °C (72–77 °F) and lows of 12–14 °C (54–57 °F). Winters are relatively cold with average high temperatures of 3 °C (37 °F) and lows of -2 to 0 °C (28 to 32 °F). Spring and autumn are generally chilly to mild. Berlin's built-up area creates a microclimate, with heat stored by the city's buildings. Temperatures can be 4 °C (7 °F) higher in the city than in the surrounding areas.[34]

The outskirts of Berlin are covered with woodlands and numerous lakes

Annual precipitation is 570 millimeters (22 in) with moderate rainfall throughout the year. Light snowfall mainly occurs from December through March, but snow cover does not usually remain for long. The recent winter of 2009/2010 was an exception as there was a permanent snow cover from late December till early March.[35]

Cityscape

Berlin along the Spree river and the Fernsehturm by night

Berlin's history has left the city with a highly eclectic array of architecture and buildings. The city's appearance today is predominantly shaped by the key role it played in Germany's history in the 20th century. Each of the national governments based in Berlin—the 1871 German Empire, the Weimar Republic, Nazi Germany, East Germany, and now the reunified Germany—initiated ambitious (re-) construction programs, with each adding its own distinctive style to the city's architecture. Berlin was devastated by bombing raids during World War II, and many of the buildings that had remained after the war were demolished in the 1950s and 1960s in both West and East Berlin. Much of this demolition was initiated by municipal architecture programs to build new residential or business quarters and main roads.

The eastern parts of Berlin have many *Plattenbauten*, reminders of Eastern Bloc ambitions to create complete residential areas that had fixed ratios of shops, kindergartens and schools to the number of inhabitants.

Architecture

The Neue Nationalgalerie by Mies van der Rohe

The Fernsehturm (TV tower) at Alexanderplatz in Mitte is among the tallest structures in the European Union at 368 meters (1207 ft). Built in 1969, it is visible throughout most of the central districts of Berlin. The city can be viewed from its 204 m (669 ft) high observation floor. Starting here the Karl-Marx-Allee heads east, an avenue lined by monumental residential buildings, designed in the Socialist Classicism Style of the Joseph Stalin era. Adjacent to this area is the Rotes Rathaus (City Hall), with its distinctive red-brick architecture. In front of it is the Neptunbrunnen, a fountain featuring a mythological group of Tritons, personifications of the four main Prussian rivers and Neptun on top of it.

The Brandenburg Gate.

The East Side Gallery is an open-air exhibition of art painted directly on the last existing portions of the Berlin Wall. It is the largest remaining evidence of the city's historical division. It has recently undergone a restoration.

The Brandenburg Gate is an iconic landmark of Berlin and Germany. It also appears on German euro coins (10 cent, 20 cent, and 50 cent). The Reichstag building is the traditional seat of the German Parliament, renovated in the 1950s after severe World War II damage. The building was again remodeled by British architect Norman Foster in the 1990s and features a glass dome over the session area, which allows free public access to the parliamentary proceedings and magnificent views of the city.

Potsdamer Platz at night.

The Gendarmenmarkt, a neoclassical square in Berlin whose name dates back to the quarters of the famous Gens d'armes regiment located here in the 18th century, is bordered by two similarly designed cathedrals, the Französischer Dom with its observation platform and the Deutscher Dom. The Konzerthaus (Concert Hall), home of the Berlin Symphony Orchestra, stands between the two cathedrals.

The Museum Island in the River Spree houses five museums build from 1830 to 1930 and is a UNESCO World Heritage site. Restoration and the construction of a main entrance to all museums, as well as the reconstruction of the Stadtschloss on the same island[37] is costing over 2 billion Euros since reunification.[38] Also located on the island and adjacent to the Lustgarten and palace is Berlin Cathedral, emperor William II's ambitious attempt to create a Protestant counterpart to St. Peter's Basilica in Rome. A large crypt houses the remains of some of the earlier Prussian royal family. The church is now owned by the Protestant umbrella Union of Evangelical Churches (UEK). Like many other buildings, it suffered extensive damage during the Second World War and had to be restored. Berlin's best preserved medieval Church of St. Mary's is the 1st preaching venue – Memorial Church being the 2nd – of the Bishop of the Evangelical Church of Berlin-Brandenburg-Silesian Upper Lusatia (EKBO), a Protestant regional church body. St. Hedwig's Cathedral is Berlin's Roman Catholic cathedral.

Kaiser Wilhelm Memorial Church is the 2nd preaching venue of the Bishop of the Regional Protestant Church (EKBO).

Unter den Linden is a tree lined east-west avenue from the Brandenburg Gate to the site of the former Berliner Stadtschloss, and was once Berlin's premier promenade. Many Classical buildings line the street and part of Humboldt University is located there. Friedrichstraße was Berlin's legendary street during the Roaring Twenties. It combines 20th century traditions with the modern architecture of today's Berlin.

Potsdamer Platz is an entire quarter built from scratch after 1995 after the Wall came down.[39] To the west of Potsdamer Platz is the Kulturforum, which houses the Gemäldegalerie, and is flanked by the Neue Nationalgalerie and the Berliner Philharmonie. The Memorial to the Murdered Jews of Europe, a Holocaust memorial, is situated to the north.[40]

The area around Hackescher Markt is home to the fashionable culture, with countless clothing outlets, clubs, bars, and galleries. This includes the Hackesche Höfe, a conglomeration of buildings around several courtyards, reconstructed around 1996. Oranienburger Straße and the nearby New Synagogue were the center of Jewish culture before 1933. Although the New Synagogue is still an anchor for Jewish history and culture, Oranienburger straße and surrounding areas are increasingly known for the shopping and nightlife.

The Straße des 17. Juni, connecting the Brandenburg Gate and Ernst-Reuter-Platz, serves as central East-West-Axis. Its name commemorates the uprisings in East Berlin of 17 June 1953. Approximately half-way from the Brandenburg Gate is the Großer Stern, a circular traffic island on which the Siegessäule (Victory Column) is situated. This monument, built to commemorate Prussia's victories, was relocated 1938–39 from its previous position in front of the Reichstag.

Schloss Charlottenburg is the largest existing palace in Berlin.

The Kurfürstendamm is home to some of Berlin's luxurious stores with the Kaiser Wilhelm Memorial Church at its eastern end on Breitscheidplatz. The church was destroyed in the Second World War and left in ruins. Nearby on Tauentzienstraße is KaDeWe, claimed to be continental Europe's largest department store. The Rathaus Schöneberg, where John F. Kennedy made his famous "Ich bin ein Berliner!" speech, is situated in Tempelhof-Schöneberg.

West of the center, Schloss Bellevue is the residence of the German President. Schloss Charlottenburg, which was burnt out in the Second World War and largely destroyed, has been rebuilt and is the largest surviving historical palace in Berlin.

The Funkturm Berlin is a 150 m (490 ft) tall lattice radio tower at the fair area, built between 1924 and 1926. It is the only observation tower which stands on insulators, and has a restaurant 55 m (180 ft) and an observation deck 126 m (413 ft) above ground, which is reachable by a windowed elevator.

Politics

Berlin is the capital of the Federal Republic of Germany. The President of Germany, whose functions are mainly ceremonial under the German constitution, has his official residence in Schloss Bellevue.[41] Berlin is the seat of the German executive, housed in the Chancellery, the *Bundeskanzleramt*.

The Reichstag is the seat of the German parliament.

Facing the Chancellery is the Bundestag, the German Parliament, housed in the renovated Reichstag building since the government moved back to Berlin in 1998. The Bundesrat ("federal council", performing the function of an upper house) is the representation of the Federal States (*Bundesländer*) of Germany and has its seat at the former Prussian House of Lords. Though most of the ministries are seated in Berlin, some of them, as well as some minor departments, are seated in Bonn, the former capital of West Germany. Discussions to move the remaining branches continuue.[42]

City state

Mayor since 2001, Klaus Wowereit

The city and state parliament is the House of Representatives (*Abgeordnetenhaus*), which currently has 141 seats. Berlin's executive body is the Senate of Berlin (*Senat von Berlin*). The Senate of Berlin consists of the Governing Mayor (*Regierender Bürgermeister*) and up to eight senators holding ministerial positions, one of them holding the official title "Mayor" (*Bürgermeister*) as deputy to the Governing Mayor. The Social Democratic Party (SPD) and The Left (Die Linke) took control of the city government after the 2001 state election and won another term in the 2006 state election.[43] After the 2011 state election, there is a coalition with the Christian Democratic Union.

The Governing Mayor is simultaneously Lord Mayor of the city (*Oberbürgermeister der Stadt*) and Prime Minister of the Federal State (*Ministerpräsident des Bundeslandes*). The office of Berlin's Governing Mayor is in the Rotes Rathaus (Red City Hall). Since 2001 this office has been held by Klaus Wowereit of the SPD.[44]

The total annual state budget of Berlin in 2007 exceeded €20.5 ($28.7) billion including a budget surplus of €80 ($112) million. The figures indicate the first surplus in the history of the city state.[45] Due to increasing growth rates and tax revenues, the Senate of Berlin calculates an increasing budget surplus in 2008. The total budget includes an estimated amount of €5.5 ($7.7) bn, which is directly financed by either the German government or the German Bundesländer.[46] Mainly due to reunification-related expenditures, Berlin as a German state has accumulated more debt than any other city in Germany, with the most current estimate being €60 ($84)bn in December 2007.[47] In 2011, the very high level of public sector debt prompted the Stabilitätsrat von Bund und Ländern (Council for Fiscal Stability of the Federal and Local States) to declare a possible fiscal emergency for the city.[48]

Since German reunification on 3 October 1990, Berlin has been one of the three city states, together with Hamburg and Bremen, among the present 16 states of Germany.

Boroughs

Map of Berlin's twelve boroughs and their localities.

Berlin is subdivided into twelve boroughs (*Bezirke*), down from 23 boroughs before Berlin's 2001 administrative reform. Each borough contains a number of localities (*Ortsteile*), which often have historic roots in older municipalities that predate the formation of Greater Berlin on 1 October 1920 and became urbanized and incorporated into the city. Many residents strongly identify with their localities or boroughs. At present Berlin consists of 95 localities, which are commonly made up of several city neighborhoods—called *Kiez* in the Berlin dialect—representing small residential areas.

Each borough is governed by a borough council (*Bezirksamt*) consisting of five councilors (*Bezirksstadträte*) and a borough mayor (*Bezirksbürgermeister*). The borough council is elected by the borough assembly (*Bezirksverordnetenversammlung*). The boroughs of Berlin are not independent municipalities, however. The power of borough governments is limited and subordinate to the Senate of Berlin. The borough mayors form the council of mayors (*Rat der Bürgermeister*), led by the city's governing mayor, which advises the senate.

The localities have no local government bodies, and the administrative duties of the former locality representative, the *Ortsvorsteher*, were taken over by the borough mayors.

Sister cities

Berlin maintains official partnerships with 17 cities.[49] Town twinning between Berlin and other cities began with Los Angeles in 1967. East Berlin's partnerships were canceled at the time of German reunification and later partially reestablished. West Berlin's partnerships had previously been restricted to the borough level. During the Cold War era, the partnerships had reflected the different power blocs, with West Berlin partnering with capitals in the West, and East Berlin mostly partnering with cities from the Warsaw Pact and its allies.

There are several joint projects with many other cities, such as Belgrade, Copenhagen, Helsinki, Johannesburg, Shanghai, Seoul, Sofia, Sydney, and Vienna. Berlin participates in international city associations such as the Union of the Capitals of the European Union, Eurocities, Network of European Cities of Culture, Metropolis, Summit Conference of the World's Major Cities, Conference of the World's Capital Cities. Berlin's official sister cities are:[49]

• 1967 **Los Angeles**, United States	• 1991 **Budapest**, Hungary	• 1994 **Tokyo**, Japan
• 1987 **Paris**, France	• 1992 **Brussels**, Belgium	• 1994 **Buenos Aires**, Argentina
• 1988 **Madrid**, Spain	• 1993 **Jakarta**, Indonesia	• 1995 **Prague**, Czech Republic
• 1989 **Istanbul**, Turkey	• 1993 **Tashkent**, Uzbekistan	• 2000 **Windhoek**, Namibia
• 1991 **Warsaw**, Poland[50]	• 1993 **Mexico City**, Mexico	• 2000 **London**, United Kingdom
• 1991 **Moscow**, Russia	• 1994 **Beijing**, China	

Economy

In 2009, the nominal GDP of the citystate Berlin experienced a growth rate of 1.7% (−3.5% in Germany) and totaled €90.1 (~$117) billion.[51] Berlin's economy is dominated by the service sector, with around 80% of all companies doing business in services. The unemployment rate reached a 15-year low in September 2011 and stood at 12.7% (German average: 6.6%).[52]

The economy of Berlin is dominated by the service sector. The ICC is part of the city's exhibition and congress center.

Fast-growing economic sectors in Berlin include communications, life sciences, and transportation, particularly services that use information and communication technologies, as well as media and music, advertising and design, biotechnology, environmental services, and medical engineering.[53]

The Science and Business Park of Berlin-Adlershof is among the 15 largest technology parks worldwide. Research and development have high economic significance for the city, and the Berlin–Brandenburg region ranks among the top-three innovative regions in the EU.[54]

2007 EUROSTAT[55]	Area		Population	Nominal GDP in billion	Nominal GDP per capita
Berlin	892 km^2	344 sq mi	3,420,000	€ 85 / ~$110	€ 24,900 / ~$32,370
Germany	357050 km^2	sq mi	82,000,000	€ 2,482 / ~$3,227	€ 29,500 / ~$38,350
EU27	4325675 km^2	sq mi	498,000,000	€ 12,363 / ~$16,072	€ 24,900 / ~$32,370

Companies

Siemens, a Fortune Global 500 company and one of the 30 German DAX companies, has a headquarter in Berlin. The state-owned railway, Deutsche Bahn, has its headquarters in Berlin as well.[56] Many German and international companies have business or service centres in the city.

Air Berlin is headquartered in Berlin.

Among the 20 largest employers in Berlin are the Deutsche Bahn, the hospital provider, Charité, the local public transport provider, BVG, and the service provider, *Dussmann and the Piepenbrock Group.* Daimler manufactures cars, and BMW builds motorcycles in Berlin. Bayer Health Care and *Berlin Chemie* are major pharmaceutical companies headquartered in the city. The second largest German airline Air Berlin is also headquartered in Berlin.[57]

Tourism

Berlin has 746 hotels with 112,400 beds as of the end of 2010. The city recorded 20.8 million overnight hotel stays and 9.1 million hotel guests in the same year.[18] Berlin has a yearly total of approximately 135 million daily visitors, which puts it in third place among the most-visited city destinations in the European Union.

Berlin is among the top three convention cities in the world and is home to Europe's biggest convention center, the Internationales Congress Centrum (ICC).[16] Several large scale trade fairs like the IFA, Grüne Woche, InnoTrans, Artforum and the ITB are held annually in the city, attracting a significant number of business visitors.

Creative industries

Industries that do business in the creative arts and entertainment are an important and sizable sector of the economy of Berlin. The creative arts sector comprises music, film, advertising, architecture, art, design, fashion, performing arts, publishing, R&D, software, TV, radio, and video games. Around 22,600 creative enterprises, predominantly SMEs, generated over 18,6 billion Euro in total revenue. Berlin's creative industries have contributed an estimated 20% of Berlin's gross domestic product in 2005.[58] The German headquarter of Universal Music is based in Berlin.

Infrastructure

Transport

Berlin Hauptbahnhof is the largest crossing station in Europe and has operated since 2006.

Berlin's transportation infrastructure is highly complex, providing a very diverse range of urban mobility.[59] A total of 979 bridges cross 197 kilometers of innercity waterways, 5334 kilometers (3314 mi) of roads run through Berlin, of which 73 kilometers (45 mi) are motorways ("Autobahn").[54] In 2006, 1.416 million motor vehicles were registered in the city.[60] With 358 cars per 1000 residents in 2008 (570/1000 in Germany), Berlin as a German state and as a major European city has one of the lowest numbers of cars per capita.[61]

Long-distance rail lines connect Berlin with all of the major cities of Germany and with many cities in neighboring European countries. Regional rail lines provide access to the surrounding regions of Brandenburg and to the Baltic Sea. The Berlin Hauptbahnhof is the largest crossing station in Europe.[62] Deutsche Bahn runs trains to domestic destinations like Hamburg, Munich, Cologne and others. It also runs an airport express rail service, as well as trains to international destinations like Moscow, Vienna, Amsterdam, and Malmö.

The Berliner Verkehrsbetriebe and the Deutsche Bahn manage several dense urban public transport systems.[63]

System	Stations/ Lines/ Net length	Passengers per year	Operator/ Notes
S-Bahn	166 / 15 / 331 km (206 mi)	376 million	DB/ Mainly overground rail system. Some suburban stops.
U-Bahn	173 / 10 / 147 km (91 mi)	457 million	BVG/ Mainly underground rail system. 24hour-service on weekends.
Tram	398 / 22 / 192 km (119 mi)	171 million	BVG/ Operates predominantly in eastern boroughs.
Bus	2627 / 147 / 1626 km (1010 mi)	407 million	BVG/ Extensive services in all boroughs. 46 Night Lines
Ferry	6 lines		BVG/ All modes of transport can be accessed with the same ticket.[64]

Airports

Tegel International Airport (left) and the Berlin Brandenburg Airport under construction (right)

Berlin has two commercial airports. Tegel Airport (TXL), which lies within the city limits, and Schönefeld Airport (SXF), which is situated just outside Berlin's south-eastern border in the state of Brandenburg. Both airports together handled 22,3 million passengers in 2010. In 2011, 88 airlines serve 164 destinations in 54 countries from Berlin.[65] Tegel Airport is the European hub of Air Berlin, whereas Schönefeld services mainly low-cost airline travel.

Berlin's airport authority plans to transfer all of Berlin's air traffic in June 2012 to a newly built airport at Schönefeld, to be renamed Berlin Brandenburg Airport (BER).[66] City authorities want to establish a European aviation hub with a gateway to Asia.

Cycling

Berlin is well known for its highly developed bike (cycle) lane system.[67] It is estimated that Berlin has 710 bicycles per 1000 residents. Around 500,000 daily bike riders accounted for 13% of total traffic in 2009.[68] Riders have access to 620 km (390 mi) of bike paths including approx. 150 km (93 mi) mandatory bicycle paths, 190 km (120 mi) off-road bicycle routes, 60 km (37 mi) of bike lanes on the roads, 70 km (43 mi) of shared bus lanes which are also open to bicyclists, 100 km (62 mi) of combined pedestrian/bike paths and 50 km (31 mi) of marked bike lanes on the sidewalks.[69]

Energy

Heizkraftwerk Mitte

Berlin's energy is mainly supplied by the Swedish firm Vattenfall, which relies more heavily than other electricity producers on lignite as an energy source. Because burning lignite produces harmful emissions, Vattenfall has announced its commitment to transitioning to cleaner sources, such as renewable energy.[70] In the former West Berlin, electricity was supplied chiefly by thermal power stations. To facilitate buffering during load peaks, accumulators were installed during the 1980s at some of these power stations. These were connected by static inverters to the power grid and were loaded during times of low energy consumption and unloaded during periods of high consumption.

In 1993 the power grid connections to the surrounding areas, which had been cut in 1951, were restored. In the western districts of Berlin, nearly all power lines are underground cables; only a 380 kV and a 110 kV line, which run from Reuter substation to the urban Autobahn, use overhead lines. The Berlin 380-kV electric line was built when West Berlin's electrical grid was not connected to those of East or West Germany. This has now become the backbone of the city's energy grid.

Car maker Daimler AG and the electric utility, RWE AG, are going to begin a joint electric car and charging station test project in Berlin called "E-Mobility Berlin."[71]

Health

The main entrance to the Charité medical campus.

Berlin has a rich history of discoveries in medicine and innovations in medical technology.[72] The modern history of medicine has been significantly influenced by scientists from Berlin. Rudolf Virchow was the founder of cellular pathology, while Robert Koch developed vaccines for anthrax, cholera, and tuberculosis.[73]

The Charité hospital complex is the largest university hospital in Europe, tracing back its origins to the year 1710. The Charité is spread over four sites and comprises 3,300 beds, around 14,000 staff, 8,000 students, and more than 60 operating theatres, and has a turnover of over one billion euros annually.[74] It is a joint institution of the Free University of Berlin and the Humboldt University of Berlin, including a wide range of institutes and specialized medical centers.

Among them are the German Heart Center, one of the most renowned transplantation centers, the Max-Delbrück-Center for Molecular Medicine and the Max-Planck-Institute for Molecular Genetics. The scientific research at these institutions is complemented by many research departments of companies such as Siemens, Schering and Debis.

Demographics

As of March 2010, the city-state of Berlin had a population of 3,440,441 registered inhabitants[1] in an area of 891.82 square kilometers (344.33 sq mi).[64] The city's population density was 3,848 inhabitants per km² (9,966/sq mi). The urban area of Berlin stretches beyond the city limits and comprises about 3.7 million people, while the metropolitan area of the Berlin-Brandenburg region is home to about 4.3 million in an area of 5370 km^2 (2070 sq mi). In 2004, The Larger Urban Zone was home to over 4.9 million people in an area of 17,385 km².[6]

Berlin's population 1880–2007.

National and international migration into the city has a long history. In 1685, following the revocation of the Edict of Nantes in France, the city responded with the Edict of Potsdam, which guaranteed religious freedom and tax-free status to French Huguenot refugees for ten years. The Greater Berlin Act in 1920 incorporated many suburbs and surrounding cities of Berlin. It formed most of the territory that comprises modern Berlin. The act increased the area of Berlin from 66 km^2 (25 sq mi) to 883 km^2 (341 sq mi) and the population from 1.9 million to 4 million.
Active immigration and asylum politics in West Berlin triggered waves of immigration in the 1960s and 1970s. Currently, Berlin is home to about 250,000 Turks (especially in Kreuzberg, Neukölln and Wedding, a locality in the borough of Mitte),[75] making it the largest Turkish community outside of Turkey.

In the 1990s the *Aussiedlergesetze* enabled immigration to Germany of some residents from the former Soviet Union. Today ethnic Germans from countries of the former Soviet Union make up the largest portion of the Russian-speaking community.[76] The current decade experiences an increasing influx from various Western countries. Especially young EU-Europeans are settling in the city. Additionally, Berlin has seen a rise of African immigrants during the last two decades.[77]

In December 2010, 457,806 residents (13.5% of the population) were of foreign nationality, originating from 190 different countries.[78] The largest groups of foreign nationals are those from Turkey (104,556), Poland (40,988), Serbia (19,230), Italy (15,842), Russia (15,332), France (13,262), Vietnam (13,199), the United States (12,733), Bosnia and Herzegovina (10,198), the United Kingdom (10,191), Croatia (10,104), and Israel (estimated 10,000[79]) .[78] An estimated 394,000 citizens (12.2%) are descendants of international migrants and have either become naturalized German citizens or obtained citizenship by virtue of birth in Germany.[80] All in all, about 25%–30% of the population is of foreign origin[81]

As of 2010, there were approx. 900,000 (approx. 27%) persons with a migrant background resident in Berlin. However, there are significant differences in the distribution of minorities. For Instance, in the West-Berlin areas of Wedding, Neukölln and Berlin-Gesundbrunnen, foreign nationals and German nationals with a migrant background make up nearly 70% of the population, whereas areas and localities in former East Berlin have much lower percentages. The immigrant community is quite diverse, however, Middle Easterners (e.g. Turks, Arabs etc.), Eastern Europeans and smaller numbers of East Asians, Sub-Saharan Africans and other European immigrants form the largest groups.[82] [83] Approximately 70,000 Afro-Germans live in Berlin.[84]

Percentage of people with migrant background[85]	
Germans without migrant background	**~73 % (2,500,000)**
Germans with migrant background (including non-German nationals)	**~27 % (900,000)**
Muslim/Middle Eastern origin (Turkey, Arab League, Iran etc.)	~9 % (300,000)
Non-German European origin (Russia, Poland, Great Britain, Greece, Serbia, Spain, France etc.)	~9 % (300,000)
Others(East Asians, Afro-Germans, Americans, Israelis, Sub-Saharan Africans, Latin Americans etc.)	~9 % (300,000)

This list is based on official statistics and not on ethnicity; hence, there might be a lower percentage of Germans without a migrant background/ethnic Germans. The percentage of children and teenagers who have a migrant background is 50%.[86] In Neukölln it is nearly 80%.[87]

Additionally, Berlin has up to 100,000 to 250,000 illegal immigrants mostly from Africa, Asia, the Balkan region and Latin America.[88] Following the accession of Romania and Bulgaria to the European Union, there has been an influx of Romani people. Local social welfare offices are attempting to integrate them and other migrants with German-language and job-skill courses.[89]

The most common foreign languages in Berlin are Turkish, Russian, Arabic, Polish, Kurdish, Vietnamese, English, Serbian, Croatian, Greek and other Asian languages. Turkish, Arabic, Kurdish, Serbian and Croatian can be heard more often in the western part, due to the large Middle-Eastern and ex-Yugoslavian immigrant communities, whereas Vietnamese, Russian and Polish have more native speakers residing in the eastern part of Berlin.[90]

Religion

More than 60% of Berlin residents have no registered religious affiliation and Berlin has been described as the *atheist capital of Europe*.[91] The largest denominations are the Protestant regional church body of the Evangelical Church of Berlin-Brandenburg-Silesian Upper Lusatia (EKBO) (a church of united administration comprising mostly Lutheran, and few Reformed and United Protestant congregations; EKBO is a member of the umbrellas Evangelical Church in Germany (EKD) and Union Evangelischer Kirchen (UEK)) with 19.4% of the population as of 2008, and the Roman Catholic Church with 9.4% of registered members.[92] [93] [94] About 2.7% of the population identify with other Christian denominations and 8.9% are Muslims. Approximately 80% of the 12,000 Jews now residing in Berlin[95] have come from the former Soviet Union.[96]

Berliner Dom, held by a congregation and the Protestant umbrella UEK.

Berlin is seat of the Roman Catholic archbishop of Berlin and also EKBO's elected chairperson is titled bishop of EKBO. Furthermore Berlin is seat of Orthodox cathedrals, such as the Cathedral of St. Boris the Baptist, one of the two seats of the Bulgarian Orthodox Diocese of Western and Central Europe, and the Resurrection of Christ Cathedral of the Diocese of Berlin (Patriarchate of Moscow).

The faithful of the different religions and denominations maintain many places of worship in Berlin. The Independent Evangelical Lutheran Church has eight parishes of different sizes in Berlin.[97] There are 36 Baptist congregations (within Union of Evangelical Free Church Congregations in Germany), 29 New Apostolic Churches, 15 United Methodist churches, eight Free Evangelical Congregations, six congregations of the Church of Jesus Christ of Latter-day Saints, an Old Catholic church and an Anglican church in Berlin. Berlin has 76 mosques, eleven synagogues, and two Buddhist temples. There are also a number of humanist and atheist groups in the city.

Education

Berlin has 878 schools that teach 340,658 children in 13,727 classes and 56,787 trainees in businesses and elsewhere.[54] The city has a six-year primary education program. After completing primary school, students progress to the Sekundarschule (a comprehensive school) or Gymnasium (college preparatory school). Berlin has a special bilingual school program embedded in the "Europaschule". At participating schools, children are taught the curriculum in German and also in a foreign language, starting in primary school and continuing in high school. Throughout nearly all boroughs, nine major European languages can be chosen as foreign languages in 29 schools.[98]

Statue of Alexander von Humboldt outside the Humboldt University

The Französisches Gymnasium Berlin, which was founded in 1689 to teach the children of Huguenot refugees, offers (German/French) instruction.[99] The John F. Kennedy School, a bilingual German–American public school located in Zehlendorf, is particularly popular with children of diplomats and the English-speaking expatriate community. In addition, four schools ("Humanistische Gymnasien") teach Latin and Classical Greek, and are renowned for highest academic standards. Two of them are state schools (Steglitzer Gymnasium in Steglitz and Goethe-Gymnasium in Wilmersdorf), one is Protestant (Evangelisches Gymnasium zum Grauen Kloster in Wilmersdorf), and one is Jesuit (Canisius-Kolleg in the "Embassy Quarter" in Tiergarten).

Science

Solar filling station at the science and technology park in Adlershof.

The Berlin-Brandenburg capital region is one of the most prolific centers of higher education and research in the European Union. The city has four universities and 27 private, professional and technical colleges (Hochschulen), offering a wide range of disciplines.[100] 135,327 students were registered at the 31 universities and colleges in 2008/09.[101] The three largest universities combined have approximately 100,000 enrolled students. They are the Humboldt Universität zu Berlin with 35,000 students, the Freie Universität Berlin (Free University of Berlin) with ca. 35,000 students, and the Technische Universität Berlin with 30,000 students. The Universität der Künste has about 4,300 students.

The city has a high density of research institutions, such as the Fraunhofer Society, Gottfried Wilhelm Leibniz Scientific Community and the Max Planck Society, which are independent of, or only loosely connected to its universities. A total number of 62,000 scientists are working in research and development.[54] The city is one of the centers of knowledge and innovation communities (Future Information and Communication Society and Climate Change Mitigation and Adaptation) of the European Institute of Innovation and Technology (EIT).[102]

In addition to libraries that are affiliated with the various universities, the Staatsbibliothek zu Berlin is a major research library. Its two main locations are near Potsdamer Platz on Potsdamer Straße and on Unter den Linden. There are also 108 public libraries in the city.[54]

Culture

The Museum Island is a World Heritage Site.

Berlin is noted for its numerous cultural institutions, many of which enjoy international reputation.[19] [103] The diversity and vivacity of the *Zeitgeist Metropolis* led to a trendsetting image among major cities.[104] The city has a very diverse art scene and is home to around 420 art galleries.[105] Many young people and international artists continue to settle in the city, and Berlin has established itself as a center of youth and popular culture in Europe.[106]

The expanding cultural role of Berlin is underscored by the 2003 announcement that the Popkomm, Europe's largest annual music industry convention—previously hosted for 15 years by Cologne—would move to Berlin.[107] Shortly thereafter, the Universal Music Group and MTV also decided to move their European headquarters and main studios to the banks of the River Spree in Friedrichshain.[108] In 2005, Berlin was awarded the title of "City of Design" by UNESCO.[17]

Media

Headquarters of the Axel Springer AG

Berlin is home to many international and regional television and radio stations.[109] The public broadcaster RBB has its headquarters in Berlin as well as the commercial broadcasters MTV Europe, VIVA, and N24. German international public broadcaster Deutsche Welle has its TV production unit in Berlin, and most national German broadcasters have a studio in the city. American radio programming from National Public Radio is also broadcast on the FM dial.

The Berlinale is the largest publicly attended film festival worldwide.

Berlin has Germany's largest number of daily newspapers, with numerous local broadsheets (*Berliner Morgenpost*, *Berliner Zeitung*, *Der Tagesspiegel*), and three major tabloids, as well as national dailies of varying sizes, each with a different political affiliation, such as *Die Welt*, *Junge Welt*, *Neues Deutschland*, and *Die Tageszeitung*. The *Exberliner*, a monthly magazine, is Berlin's English-language periodical focusing on arts and entertainment. Berlin is also the headquarter of the two major German-language publishing houses Walter de Gruyter and Springer, each of which publishing books, periodicals, and multimedia products.

Berlin is an important center in the European and German film industry.[110] It is home to more than 1000 film and television production companies, 270 movie theaters, and around 300 national and international co-productions are filmed in the region every year.[54] The historic Babelsberg Studios and the production company UFA are located outside Berlin in Potsdam. The city is also home of the

The Buddy Bear: A Symbol of *New Berlin*

European Film Academy and the German Film Academy, and hosts the annual Berlin Film Festival. Founded in 1951, the festival has been celebrated annually in February since 1978. With over 430,000 admissions it is the largest publicly attended film festival in the world.[111]

Nightlife and festivals

Berghain

Berlin's nightlife is one of the most diverse and vibrant of its kind in Europe.[112] Throughout the 1990s, people in their twenties from many countries, particularly those in Eastern and Central Europe, made Berlin's club scene the premier nightlife destination of Europe. After the fall of the Berlin Wall in 1989, many historic buildings in Mitte, the former city center of East Berlin, were illegally occupied and re-built by young squatters and became a fertile ground for underground and counterculture gatherings. Mitte and surrounding boroughs are also home to many nightclubs, including Kunst Haus Tacheles, techno clubs Tresor, WMF, Ufo, E-Werk, KitKatClub and Berghain. The techno-music club, Linientreu, near the Kaiser Wilhelm Memorial Church, has been in business since the late 1980s. The LaBelle discothèque in Friedenau became widely known as the location of the 1986 Berlin discotheque bombing.[113] Berlin is notable for the length of its parties. Clubs are not required to close at a fixed time on the weekends, and many parties last well into the morning, or all weekend. Berghain features the Panorama Bar, so named because the bar opens its shades at daybreak, allowing party-goers a panorama view of Berlin after dancing through the night.

Beach along the Spree river

The SO36 in Kreuzberg originally focused largely on punk music, but today has become a popular venue for many dances and parties. SOUND, located from 1971 to 1988 in Tiergarten and today in Charlottenburg, gained notoriety in the late 1970s for its popularity with heroin users and other drug addicts as described in Christiane F.'s book *Wir Kinder vom Bahnhof Zoo.*[114]

The Karneval der Kulturen, a multi-ethnic street parade celebrated every Pentecost weekend,[115] and the Christopher Street Day are both supported by the city's government.[116] Berlin is also well known for the cultural festival, Berliner Festspiele, which include the jazz festival JazzFest Berlin. Several technology and media art festivals and conferences are held in the city, including Transmediale and Chaos Communication Congress.

Gay life

Berlin has a long history of gay culture and influence on popular entertainment, and according to some authors, in the 1920s the city was the *Gay Capital of Europe*.[117] Today, the city has a huge number of gay clubs and festivals, such as Easter fetish week (Easter in Berlin), Christopher Street Day (Berlin Pride)—central Europe's largest gay-lesbian pride event celebrated on the last weekend of June—Folsom Europe and Hustlaball. Berlin is also leading Europe in the number of fetish clubs. "Easter in Berlin" and "Folsom Europe Berlin" are the biggest gay fetish festivals in Europe. Annual gay highlights in Berlin are also the gay and lesbian street festival in Berlin-Schöneberg (Lesbisch-schwules Stadtfest) and Kreuzberg Pride in June. The largest gay areas in Berlin are located in Schöneberg close to Nollendorfplatz and in Prenzlauer Berg at the Schönhauser Allee subway station.[118][119]

Galleries and museums

Berlin is home to 153 museums.[54] The ensemble on the Museum Island is a UNESCO World Heritage Site and is situated in the northern part of the Spree Island between the Spree and the Kupfergraben.[19] As early as 1841 it was designated a "district dedicated to art and antiquities" by a royal decree. Subsequently, the Altes Museum (Old Museum) in the Lustgarten displaying the bust of Queen Nefertiti,[120] the Neues Museum (New Museum), Alte Nationalgalerie (Old National Gallery), Pergamon Museum, and Bode Museum were built there. While these buildings once housed distinct collections, the names of the buildings no longer necessarily correspond to the names of their collections.

The Jewish Museum presents an exhibition on two millennia of German–Jewish history.

Apart from the Museum Island, there are many additional museums in the city. The Gemäldegalerie (Painting Gallery) focuses on the paintings of the "old masters" from the 13th to the 18th centuries, while the Neue Nationalgalerie (New National Gallery, built by Ludwig Mies van der Rohe) specializes in 20th century European painting. The Hamburger Bahnhof, located in Moabit, exhibits a major collection of modern and contemporary art. In spring 2006, the expanded Deutsches Historisches Museum re-opened in the Zeughaus with an overview of German history through the fall of the Berlin Wall in 1989. The Bauhaus Archive is an architecture museum.

The reconstructed Ishtar Gate of Babylon at the Pergamon Museum.

The Jewish Museum has a standing exhibition on two millennia of German-Jewish history.[121] The German Museum of Technology in Kreuzberg has a large collection of historical technical artifacts. The Museum für Naturkunde exhibits natural history near Berlin Hauptbahnhof. It has the largest mounted dinosaur in the world (a brachiosaurus), and a preserved specimen of the early bird Archaeopteryx.[122]

In Dahlem, there are several museums of world art and culture, such as the Museum of Asian Art, the Ethnological Museum, the Museum of European Cultures, as well as the Allied Museum (a museum of the Cold War) and the Brücke Museum (an art museum). In Lichtenberg, on the grounds of the former East German Ministry for State Security *(Stasi)*, is the Stasi Museum. The site of Checkpoint Charlie, one of the most renowned crossing points of the Berlin Wall, is still preserved and also has a museum, a private venture which exhibits comprehensive documentation of detailed plans and strategies devised by people who tried to flee from the East. The Beate Uhse Erotic Museum near Zoo Station claims to be the world's largest erotic museum.[123]

Performing arts

Sir Simon Rattle conducting the renowned Berlin Philharmonic.

Berlin is home to more than 50 theaters.[54] The Deutsches Theater in Mitte was built in 1849–50 and has operated continuously since then, except for a one-year break (1944–45) due to the Second World War. The Volksbühne at Rosa Luxemburg Platz was built in 1913–14, though the company had been founded in 1890. The Berliner Ensemble, famous for performing the works of Bertolt Brecht, was established in 1949, not far from the Deutsches Theater. The Schaubühne was founded in 1962 in a building in Kreuzberg, but in 1981 moved to the building of the former Universum Cinema on Kurfürstendamm.

Haus der Kulturen der Welt

Berlin has three major opera houses: the Deutsche Oper, the Berlin State Opera, and the Komische Oper. The Berlin State Opera on Unter den Linden opened in 1742 and is the oldest of the three. Its current musical director is Daniel Barenboim. The Komische Oper has traditionally specialized in operettas and is located at Unter den Linden as well. The Deutsche Oper opened in 1912 in Charlottenburg. During the division of the city from 1961 to 1989 it was the only major opera house in West Berlin.

There are seven symphony orchestras in Berlin. The Berlin Philharmonic Orchestra is one of the preeminent orchestras in the world;[124] it is housed in the Berliner Philharmonie near Potsdamer Platz on a street named for the orchestra's longest-serving conductor, Herbert von Karajan.[125] The current principal conductor is Simon Rattle.[126] The Konzerthausorchester Berlin was founded in 1952 as the orchestra for East Berlin, since the Philharmonic was based in West Berlin. Its current principal conductor is Lothar Zagrosek. The Haus der Kulturen der Welt presents various exhibitions dealing with intercultural issues and stages world music and conferences.[127]

Cuisine

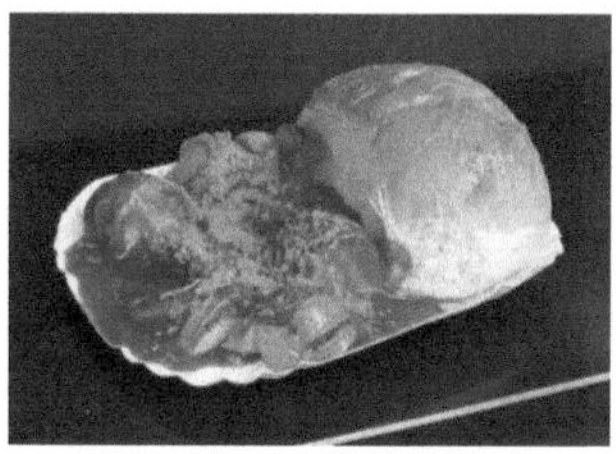

The currywurst was invented in Berlin

Berlin is home to a diverse gastronomy scene reflecting the immigrant history of the city. Twelve restaurants in Berlin have been included into the Michelin guide, which ranks the city at the top for the number of its restaurants having this distinction in Germany.[128]

Many local foods originated from north-German culinary traditions and include rustic and hearty dishes with pork, goose, fish, peas, beans, cucumbers or potatoes.

Typical Berliner fares include Currywurst, invented in 1949,[129] Eisbein, the Berliner known as a *Pfannkuchen*, and *Leber Berliner Art* (Berlin-style liver).[130]

Turkish and Arab immigrant workers brought their culinary traditions to the city; for example, the döner kebab, falafel and lahmacun, which have become common fast-food staples. The modern fast-food version of the döner was invented in Berlin in 1971.[131]

Recreation

The Zoologischer Garten Berlin is the most visited zoo in Europe and presents the most diverse range of species in the world.

Zoologischer Garten Berlin, the older of two zoos in the city, was founded in 1844, and presents the most diverse range of species in the world.[132] It was the home of the captive-born celebrity polar bear Knut,[133] born in December 2006.[134] The city's other zoo is Tierpark Friedrichsfelde, founded in 1955 on the grounds of Schloss Friedrichsfelde in the Borough of Lichtenberg.

Berlin's Botanischer Garten includes the Botanic Museum Berlin. With an area of 43 hectares (110 acres) and around 22,000 different plant species it is one of the largest and most diverse gardens in the world. Other gardens in the city include the Britzer Garten, site of the 1985 Bundesgartenschau, and the Erholungspark Marzahn, promoted under the name *Gardens of the world*.[135]

The Tiergarten is Berlin's largest park located in Mitte and was designed by Peter Joseph Lenné.[136] In Kreuzberg the Viktoriapark provides a good viewing point over the southern part of inner city Berlin. Treptower Park beside the Spree in Treptow has a monument honoring the Soviet soldiers killed in the 1945 Battle of Berlin. The Volkspark in Friedrichshain, which opened in 1848, is the oldest park in the city. Its summit is man-made and covers a Second World War bunker and rubble from the ruins of the city; at its foot is Germany's main memorial to Polish soldiers.

Berlin is known for its numerous beach bars along the river Spree. Together with the countless cafés, restaurants and green spaces in all districts, they create an important source of recreation and leisure time.[137]

Sports

Berlin has established a high-profile reputation as a host city of international sporting events.[138] Berlin hosted the 1936 Olympics and was the host city for the 2006 FIFA World Cup Final.[139] The IAAF World Championships in Athletics were held in the Olympiastadion in August 2009.[140] The annual Berlin Marathon and the annual ÅF Golden League event ISTAF for athletics are also held here.[141] The FIVB World Tour has chosen an inner-city site near Alexanderplatz to present a beach volleyball Grand Slam every year.

The Olympiastadion hosted the 1936 Summer Olympics and the 2006 FIFA World Cup final.

Open Air gatherings of several hundred thousands spectators have become popular during international football competitions, like the World Cup or the UEFA European Football Championship. Many fans and viewers are coming together to watch the matches on huge video screens. The event is known as the *Fan Mile* and takes place at the Brandenburg Gate every two years.[142]

Several major clubs representing the most popular spectator sports in Germany have their base in Berlin.

The annual Berlin Marathon is known as a flat and fast course.

Club	Sport	Founded	League	Venue	Head Coach
Hertha BSC[143]	Football	1892	Bundesliga	Olympiastadion	M. Babbel
1. FC Union Berlin[144]	Football	1966	2. Bundesliga	Alte Försterei	U. Neuhaus
ALBA Berlin[145]	Basketball	1991	BBL	O_2 World	Gordon Herbert
Eisbären Berlin[146]	Ice hockey	1954	DEL	O_2 World	D. Jackson
Füchse Berlin[147]	Handball	1891	HBL	Max-Schmeling-Halle	D. Sigurdsson
Berlin Recycling Volleys[148]	Volleyball	1911	DVL[149]	Max-Schmeling-Halle	Mark Lebedew

Quotations

Marlene Dietrich was born in Berlin-Schöneberg

- "Berlin ist eine Stadt, verdammt dazu, ewig zu werden, niemals zu sein" ("Berlin is a city condemned always to become, never to be.")
  (Karl Scheffler, author of *Berlin: Ein Stadtschicksal, 1910*)[150]
- "Ich hab noch einen Koffer in Berlin" ("I still have a suitcase in Berlin")
  (Marlene Dietrich, 1951 song by the actress and singer born in Berlin-Schöneberg.)[151]
- "Ich bin ein Berliner". ("I am a citizen of Berlin")
  (John F. Kennedy, President of the United States, 1963 while visiting Berlin)
- "The greatest cultural extravaganza that one could imagine."
  (David Bowie, singer, on 1970s Berlin)[152]
- "Berlin wird leben und die Mauer wird fallen." ("Berlin will live and the wall will fall.")
  (Willy Brandt, Former Governing Mayor of West Berlin and chancellor of Germany, 10 November 1989)[153]
- "Berlin ist arm, aber sexy." ("Berlin is poor, but sexy.")
  (Klaus Wowereit, Governing Mayor, in a press interview, 2003)[154]

Notes

[1] "Bevölkerungsstand in Berlin am 31. Juli 2011 nach Bezirken" (http://www.statistik-berlin-brandenburg.de//Publikationen/OTab/2011/OT_A01-01-00_124_201107_BE.pdf) (in German). *Amt für Statistik Berlin-Brandenburg*. 31 July 2011. . Retrieved 14 November 2011.

[2] Prefixes for vehicle registration were introduced in 1906, but often changed due to the political changes after 1945. Vehicles were registered under the following prefixes: "I A" (1906– April 1945; devalidated on 11 August 1945); no prefix, only digits (since July till August 1945), "БГ" (=BG; 1945–1946, for cars, lorries and busses), "ГФ" (=GF; 1945–1946, for cars, lorries and busses), "БМ" (=BM; 1945–1947, for motor bikes), "ГМ" (=GM; 1945–1947, for motor bikes), "KB" (i.e.: Kommandatura of Berlin; for all of Berlin 1947–1948, continued for West Berlin until 1956), "GB" (i.e.: Greater Berlin, for East Berlin 1948–1953), "I" (for East Berlin, 1953–1990), "B" (for West Berlin as of 1 July 1956, continued for all of Berlin since 1990).

[3] "Bruttoinlandsprodukt (nominal) in BERLIN seit 1995" (http://www.berlin.de/imperia/md/content/sen-wirtschaft/konjunkturdaten/a_03.pdf?start&ts=1272613011&file=a_03.pdf) (in German). 30 March 2010. . Retrieved 15 May 2011.

[4] http://www.berlin.de/international/index.en.php

[5] INSEE. "Population des villes et unités urbaines de plus de 1 million d'habitants de l'Union européenne" (http://www.insee.fr/fr/themes/tableau.asp?ref_id=cmptef01103&id=18) (in French). . Retrieved 17 August 2008.

[6] "City Profiles Berlin" (http://www.urbanaudit.org/CityProfiles.aspx). *Urban Audit*. . Retrieved 20 August 2008.

[7] *Gren Berlin* (http://books.google.de/books?id=94LP4xCb-KcC&pg=PA23&lpg=PA23&dq="Berlin+"+rivers+lakes+foreests+"one+third"&source=bl&ots=OoT9TA0ZWR&sig=ZN0MSyam43OC5321qPitypjubUg&hl=en&ei=VWTPStrdKabEmwPdx9yEAw&sa=X&oi=book_result&ct=result&resnum=1#v=onepage&q=&f=false). . Retrieved 9 October 2009.

[8] "Documents of German Unification, 1848–1871" (http://www.fordham.edu/halsall/mod/germanunification.html). *Modern History Sourcebook*. . Retrieved 18 August 2008.

[9] "Topographies of Class: Modern Architecture and Mass Society in Weimar Berlin (Social History, Popular Culture, and Politics in Germany)." (http://www.h-net.org/reviews/showrev.php?id=23505). */www.h-net.org*. . Retrieved 9 October 2009.

[10] "Berlin Wall" (http://www.britannica.com/EBchecked/topic/62202/Berlin-Wall). *Encyclopædia Britannica*. . Retrieved 18 August 2008.

[11] "Berlin – Capital of Germany" (http://www.germany.info/Vertretung/usa/en/04__W__t__G/03/01/03/Feature__3.html). *German Embassy in Washington*. . Retrieved 18 August 2008.

[12] "Diplomatic Berlin" (http://www.magazine-deutschland.de/en/artikel-en/article/article/das-diplomatische-berlin.html). *Deutschland Online*. . Retrieved 7 March 2009.

[13] Davies, Catriona (10 April 2010). "Revealed: Cities that rule the world – and those on the rise" (http://edition.cnn.com/2010/BUSINESS/04/10/cities.dominate.world/?hpt=C2). CNN. . Retrieved 11 April 2010.

[14] Sifton, Sam (31 December 1969). "Berlin, the big canvas" (http://www.nytimes.com/2008/06/22/travel/22iht-22berlin.13882912.html?_r=1). *The New York Times*. . Retrieved 18 August 2008. See also: "Sites and situations of leading cities in cultural globalisations/Media" (http://www.lboro.ac.uk/gawc/rb/rb146.html#t5). *GaWC Research Bulletin 146*. . Retrieved 18 August 2008.

[15] "Global Power City Index 2009" (http://www.mori-m-foundation.or.jp/english/research/project/6/pdf/GPCI2009_English.pdf). *Institute for Urban Strategies at The Mori Memorial Foundation* (Tokyo, Japan). 22 October 2009. . Retrieved 29 October 2009.

[16] "ICCA publishes top 20 country and city rankings 2007" (http://www.iccaworld.com/npps/story.cfm?ID=1577). *ICCA*. . Retrieved 18 August 2008.

[17] "Berlin City of Design Press Release" (http://portal.unesco.org/culture/en/ev.php-URL_ID=29376&URL_DO=DO_TOPIC&URL_SECTION=201.html). *UNESCO*. . Retrieved 18 August 2008.

[18] "Berlin-Tourismus 2010 mit neuem Rekord" (http://www.statistik-berlin-brandenburg.de/pms/2011/11-02-18.pdf) (in German). *Amt für Statistik*. . Retrieved 19 February 2011.

[19] "World Heritage Site Museumsinsel" (http://whc.unesco.org/en/list/896). *UNESCO*. . Retrieved 18 August 2008.

[20] "Hollywood Helps Revive Berlin's Former Movie Glory" (http://www.dw-world.de/dw/article/0,2144,3549403,00.html). *Deutsche Welle*. 9 August 2008. . Retrieved 18 August 2008.

[21] Flint, Sunshine (12 December 2004). "The Club Scene, on the Edge" (http://travel2.nytimes.com/2004/12/12/travel/12surf.html?ex=1150862400&en=f2c1cc6c507fea18&ei=5070). *The New York Times*. . Retrieved 18 August 2008. See also: "Ranking of best cities in the world" (http://www.citymayors.com/features/quality_survey.html). *City mayors*. . Retrieved 18 August 2008. and "The Monocle Quality Of Life Survey" (http://www.denmark.dk/NR/rdonlyres/60B9101D-B656-438E-A755-E441D41E0AA7/0/top_20_cities.pdf) (PDF). *www.denmark.dk*. . Retrieved 18 August 2008.

[22] Berger, Dieter (1999). *Geographische Namen in Deutschland*. Bibliographisches Institut. ISBN 3411062525.

[23] "Berlin dig finds city older than thought". Associated Press.

[24] "Spandau Citadel" (http://web.archive.org/web/20080612020333/http://www.berlin.de/tourismus/sehenswuerdigkeiten.en/00175.html). *Berlin tourist board*. Archived from the original (http://www.berlin.de/tourismus/sehenswuerdigkeiten.en/00175.html) on 12 June 2008. . Retrieved 18 August 2008.

[25] "The Hohenzollern Dynasty" (http://web.archive.org/web/20070807093738/http://www.west.net/~antipas/protected_files/news/europe/hohenzollerns.html). *Antipas*. Archived from the original (http://www.west.net/~antipas/protected_files/news/europe/hohenzollerns.html) on 7 August 2007. . Retrieved 18 August 2008.

[26] "Berlin Cathedral" (http://web.archive.org/web/20060818100934/http://www.smp-protein.de/SMPConference/berlin.htm). *SMPProtein*. Archived from the original (http://www.smp-protein.de/SMPConference/berlin.htm) on 18 August 2006.. Retrieved 18 August 2008.

[27] "Brandenburg during the 30 Years War" (http://www.zum.de/whkmla/region/germany/bra30.html). *WHKMLA*.. Retrieved 18 August 2008.

[28] Clodfelter, Michael (2002), *Warfare and Armed Conflicts- A Statistical Reference to Casualty and Other Figures, 1500–2000* (2nd ed.), McFarland & Company, ISBN 0786412046

[29] "Agreement to divide Berlin" (http://www.fdrlibrary.marist.edu/psf/box31/t297c01.html). *FDR-Library*.. Retrieved 18 August 2008.

[30] "Berlin Airlift / Blockade" (http://www.western-allies-berlin.com/historic-events/detail/airlift-blockade). *Western Allies Berlin*.. Retrieved 18 August 2008.

[31] "Berlin official website; History after 1945" (http://www.berlin.de/berlin-im-ueberblick/geschichte/1945.en.html). *City of Berlin*.. Retrieved 8 April 2009.

[32] "Ostpolitik: The Quadripartite Agreement of September 3, 1971" (http://usa.usembassy.de/etexts/ga5-710903.htm). *US Berlin Embassy*. . Retrieved 18 August 2008.

[33] "Satellite Image Berlin" (http://maps.google.com/maps?ll=52.5333,13.38000&spn=0.060339,0.085316&t=k&hl=en). *Google Maps*. . Retrieved 18 August 2008.

[34] www.weather.com

[35] "Climate figures" (http://www.worldweather.org/016/c00059.htm). *World Weather Information Service*.. Retrieved 18 August 2008.

[36] "Climatological Normals of Berlin" (http://www.hko.gov.hk/wxinfo/climat/world/eng/europe/ger_pl/berlin_e.htm). *Hong Kong Observatory*.. Retrieved 20 May 2010.

[37] Neumann: Stadtschloss wird teurer, Berliner Zeitung 24 June 2011 (http://www.berliner-zeitung.de/newsticker/neumann--stadtschloss-wird-teurer,10917074,10924086.html)

[38] Das Pathos der Berliner Republik, Berliner Zeitung, 19 May 2010 (http://www.berliner-zeitung.de/archiv/die-nullerjahre--nation-building---der-wiedervereinigte-staat-baut-sich-eine-neue-hauptstadt-das-pathos-der-berliner-republik,10810590,10717494.html)

[39] "Construction and redevelopment since 1990" (http://www.stadtentwicklung.berlin.de/bauen/baubilanz/en/potsdamer_platz.html). *Senate Department of Urban Development*.. Retrieved 18 August 2008.

[40] Ouroussoff, Nicolai (9 May 2005). "A Forest of Pillars, Recalling the Unimaginable" (http://travel2.nytimes.com/2005/05/09/arts/design/09holo.html?ex=1162962000&en=272335914d8b80d6&ei=5070). *The New York Times*.. Retrieved 18 August 2008.

[41] Bundespräsident Horst Köhler (http://www.bundespraesident.de/DE/Home/home_node.html), www.bundespraesident.de. Retrieved 12 November 2006.

[42] Der Regierungsumzug ist überfällig, Berliner Zeitung, 26 October 2010 (http://www.berliner-zeitung.de/archiv/der-regierungsumzug-ist-ueberfaellig,10810590,10751012.html)

[43] "Berlin state election, 2006" (http://www.statistik-berlin.de/produkte/Faltblatt_Brochure/berlin_in_Zahlen_engl.pdf) (in German). *Der Landeswahlleiter für Berlin*.. Retrieved 17 August 2008.

[44] "The Glamor Guy" (http://www.time.com/time/world/article/0,8599,2051103,00.html). *Time Europe*. 8 May 2005.. Retrieved 17 August 2008. See also: Landler, Mark (23 September 2006). "Berlin Mayor, Symbol of Openness, Has National Appeal" (http://www.nytimes.com/2006/09/23/world/europe/23wowereit.html?ex=1161489600&en=2129daf9965bf613&ei=5070). *The New York Times*.. Retrieved 17 August 2008.

[45] Berlin schafft erstes Etatplus seit dem Krieg (http://www.spiegel.de/wirtschaft/0,1518,527721,00.html)(German), SpiegelOnline. Retrieved February, 2008.

[46] Fahrun, Joachim (10 June 2008). "Sarrazin: Keine neuen Schulden mehr ab 2008" (http://www.morgenpost.de/printarchiv/berlin/article199345/Sarrazin_Keine_neuen_Schulden_mehr_ab_2008.htm). *Berliner Morgenpost*.. Retrieved 17 August 2008.

[47] "Debt-Laden Berlin Goes to Court For Federal Aid" (http://www.dw-world.de/dw/article/0,2144,1983556,00.html). *Deutsche Welle*. 29 April 2006.. Retrieved 20 October 2006.

[48] "Vier Bundesländern drohen griechische Verhältnisse" (http://www.spiegel.de/wirtschaft/soziales/0,1518,764436,00.html) (in German). *Spiegel*.. Retrieved 23 May 2011.

[49] "Berlin's international city relations" (http://www.berlin.de/rbmskzl/staedteverbindungen/staedtepartnerschaft_ueberblick.de.html). *Berlin Mayor's Office*.. Retrieved 22 August 2010.

[50] "Miasta partnerskie Warszawy" (http://www.um.warszawa.pl/node/2920?page=0,0). *um.warszawa.pl*. Biuro Promocji Miasta. 4 May 2005.. Retrieved 29 August 2008.

[51] "Exporteinbruch stürzt deutsche Wirtschaft in die Krise" (http://de.reuters.com/article/topNews/idDEBEE51O0CG20090225) (in German). Reuters.. Retrieved 19 August 2008.

[52] "Arbeitslosigkeit sinkt auf niedrigsten Stand seit 20 Jahren" (http://www.morgenpost.de/printarchiv/titelseite/article1779972/Arbeitslosigkeit-sinkt-auf-niedrigsten-Stand-seit-20-Jahren.html) (in German). Berliner Morgenpost.. Retrieved 30 September 2011.

[53] "Poor but sexy" (http://www.economist.com/displayStory.cfm?story_id=7953479). *The Economist*. 21 September 2006.. Retrieved 19 August 2008.

[54] "Berlin fact sheet" (http://www.stadtentwicklung.berlin.de/internationales_eu/staedte_regionen/en/eurocities_2011.shtml) (PDF). *berlin.de*.. Retrieved 19 August 2008.

[55] "Regional GDP per inhabitant in the EU 27" (http://docs.google.com/viewer?a=v&q=cache:_JgilxNoYRQJ:epp.eurostat.ec.europa.eu/cache/ITY_PUBLIC/1-18022010-AP/EN/1-18022010-AP-EN.PDF+eurostat+Regional+GDP+per+inhabitant+in+the+EU+27&hl=de&gl=de&pid=bl&srcid=ADGEESimOFSZQ3D5aE2vWgMR3IFlQCzBcdl7NBIWfHjdZMvpwFW1fLwhHg2jvwx0OOnFsKVvbCm1fTNaTNtQ3K0o-8wEm_TAREoTOu sig=AHIEtbQqggVoFeopAcI9aBQ75d0RvwSVRQ) (PDF). *Eurostat*. . Retrieved 17 May 2010.

[56] "DB Schenker to concentrate control functions in Frankfurt am Main" (http://www.deutschebahn.com/site/bahn/en/group/group__affairs/schenker__frankfurt__20100714.html). . Retrieved 6 June 2011.

[57] " Air Berlin: Contact (http://www.airberlin.com/site/kontakt_db_a.php?LANG=eng)." Retrieved on 12 May 2009.

[58] Creative Industries in Berlin (http://www.kulturwirtschaft.de/?p=573) berlin.de Retrieved on 1 April 2011.

[59] "Broke but dynamic, Berlin seeks new identity" (http://www.iht.com/articles/2005/05/01/news/mayor5.php). *IHT*. . Retrieved 19 August 2008.

[60] "Kraftfahrzeuge und Schienenbestand" (http://www.destatis.de/jetspeed/portal/cms/Sites/destatis/Internet/DE/Content/Statistiken/Verkehr/VerkehrsmittelbestandInfrastruktur/Tabellen/Content75/Fahrzeugbestand,templateId=renderPrint.psml). *Destatis*. . Retrieved 19 August 2008.

[61] "Die Hauptstadt des umweltfreundlichen Verkehrs" (http://www.berlinonline.de/berliner-zeitung/archiv/.bin/dump.fcgi/2009/0811/berlin/0025/index.html) (in German). *Berliner Zeitung*. . Retrieved 11 August 2009.

[62] "Berlin Hauptbahnhof" (http://www.hbf-berlin.de/site/berlin__hauptbahnhof/en/start.html). *Deutsche Bahn*. . Retrieved 2 May 2009.

[63] Amt für Statistik Berlin-Brandenburg, Die kleine Berlin-Statistik (http://www.statistik-berlin-brandenburg.de/produkte/KleineStatistik/kBEst_2008.pdf). (German) Accessed 6 January 2009.

[64] "Berlin statistical figures" (http://www.statistik-berlin.de/) (in German). *Amt für Statistik Berlin-Brandenburg*. . Retrieved 19 August 2008.

[65] "Airport BBI: Einzigartige, historische Chance für Berlin und Brandenburg" (http://www.airportzentrale.de/?p=4940). *airportzentrale.de*. . Retrieved 10 March 2011.

[66] "Airport Berlin Brandenburg International" (http://www.berlin-airport.de/EN/BBI/). *Airports Berlin*. . Retrieved 7 July 2010.

[67] "Bike City Berlin" (http://www.treehugger.com/files/2007/07/bike_city_berli.php). *Treehugger*. . Retrieved 19 August 2008.

[68] "Platz da! – für die Radfahrer" (http://www.neues-deutschland.de/artikel/193840.platz-da-8211-fuer-die-radfahrer.html). *ND*. . Retrieved 22 March 2011.

[69] "Bicycle Routes and Facilities Bicycle Paths" (http://www.stadtentwicklung.berlin.de/verkehr/radverkehrsanlagen/en/radwege.shtml). *Senate Department of urban development*. . Retrieved 19 September 2008.

[70] Reimer, Nick (30 May 2006). "Abgase tiefer gelegt" (http://www.taz.de/index.php?id=archivseite&dig=2006/05/30/a0166) (in German). *taz*. . Retrieved 18 August 2008.

[71] "CNN.com Video" (http://edition.cnn.com/video/). *CNN*. .

[72] Kühne, Anja; Warnecke, Tilmann (17 October 2007). "Berlin leuchtet" (http://www.tagesspiegel.de/magazin/wissen/Elitewettbewerb-Universitaeten;art304,2400973) (in German). *Der Tagesspiegel*. . Retrieved 18 August 2008.

[73] "History of the Charité of Berlin" (http://www.charite.de/ch/neuro/lokales/charite/geschichte.html). *Charité*. 6 September 2005. . Retrieved 18 August 2008.

[74] "Customer company profile" (http://www.anselleurope.com/medical/index.cfm?Include=testimonials&lang=EN&I=4166). *Ansell Healthcare*. . Retrieved 18 August 2008.

[75] Spooner, Andrew (13 May 2007). "Berlin: Shish And Sauerkraut To Go" (http://www.independent.co.uk/travel/europe/berlin-shish-and-sauerkraut-to-go-448678.html). *The Independent* (UK). . Retrieved 24 May 2010.

[76] Berlin is speaking Russians' language (http://www.russiajournal.com/node/4653). The Russia Journal. 10 March 2001.

[77] "Berlin wird farbiger. Die Afrikaner kommen – Nachrichten WELT am SONNTAG – WELT ONLINE" (http://www.welt.de/print-wams/article616463/Berlin_wird_farbiger_Die_Afrikaner_kommen.html) (in (German)). *Die Welt*. 28 October 2001. . Retrieved 2 June 2011.

[78] "Foreign residents of Berlin" (http://www.statistik-berlin-brandenburg.de/pms/2011/11-02-04.pdf) (in German) (PDF). *Amt für Statistik Berlin-Brandenburg*. . Retrieved 5 February 2011.

[79] Israelis Learn to Love the New Berlin (http://www.spiegel.de/international/zeitgeist/0,1518,740410,00.html). SpiegelOnline. 5 February 2011.

[80] "Statistik Berlin/Brandbenburg" (http://www.statistik-berlin-brandenburg.de/Publikationen/Stat_Berichte/2011/SB_A1-5_hj02-10_BE.pdf). Amt für Statistik Berlin-Brandenburg. 31 December 2010. . Retrieved 1 January 2009.

[81] "Jeder vierte Berliner ist ausländischer Herkunft" (http://www.tagesspiegel.de/berlin/;art270,2563533) (in German). *Der Tagesspiegel*. . Retrieved 19 August 2008.

[82] http://www.statistik-berlin-brandenburg.de/Publikationen/Stat_Berichte/2011/SB_A1-5_hj02-10_BE.pdf

[83] "Migration - Jeder vierte Berliner hat ausländische Wurzeln - Berlin Aktuell - Berliner Morgenpost - Berlin" (http://www.morgenpost.de/berlin-aktuell/article1592839/Jeder-vierte-Berliner-hat-auslaendische-Wurzeln.html). Morgenpost.de. 2011-03-30. . Retrieved 2011-09-15.

[84] Yonis Ayeh. "ISD Online • Initiative Schwarze Menschen in Deutschland" (http://www.isdonline.de). Isdonline.de. . Retrieved 2011-09-15.

[85] http://www.statistik-berlin-brandenburg.de/pms/2011/11-02-04.pdf

[86] "Schlagworte der Integrationsdebatte" (http://www.berlin.de/lb/intmig/schlagwort_02.html). Berlin.de. . Retrieved 2011-09-15.

[87] "deine inform@tion zone: Mach dich stark für starke Kinder" (http://www.wob4u.de/1651.html?&tx_ttnews[tt_news]=912&cHash=bc9a3fcc745c0059db2bf96456755ee2). wob4u.de. . Retrieved 2011-09-15.

[88] Von Andrea Dernbach (2009-02-23). "Migration: Berlin will illegalen Einwanderern helfen - Deutschland - Politik - Tagesspiegel" (http://www.tagesspiegel.de/politik/deutschland/berlin-will-illegalen-einwanderern-helfen/1452916.html). Tagesspiegel.de. . Retrieved 2011-09-15.

[89] "Die Roma von Berlin-Neukölln - Nachrichten welt_print - Politik - WELT ONLINE" (http://www.welt.de/die-welt/politik/article9920445/Die-Roma-von-Berlin-Neukoelln.html) (in (German)). Welt.de. 2010-09-28. . Retrieved 2011-09-15.

[90] "Studie – Zwei Millionen Berliner sprechen mindestens zwei Sprachen – Wirtschaft – Berliner Morgenpost – Berlin" (http://www.morgenpost.de/printarchiv/wirtschaft/article1309952/Zwei-Millionen-Berliner-sprechen-mindestens-zwei-Sprachen.html). Morgenpost.de. 18 May 2010. . Retrieved 2 June 2011.

[91] Connolly, Kate (26 April 2009). "Atheist Berlin to decide on religion's place in its schools" (http://www.guardian.co.uk/world/2009/apr/26/berlin-germany-religious-education-ethics). *The Guardian* (UK). . Retrieved 10 May 2010.

[92] Evangelische Kirche im Rheinland, Table 1.3: Bevölkerung und Kirchenzugehörigkeit nach Bundesländern (http://www.ekir.de/ekir/dokumente/Tab01.pdf). Retrieved 13 April 2010.

[93] Evangelical Church in Germany, Kirchenmitglieder 2007 (http://www.ekd.de/download/kirchenmitglieder_2007.pdf). Tabelle 2: Evangelische Kirchenmitglieder und Bevölkerung nach Bundesländern am 31 December 2007 and also Tabelle 3: Evangelische Kirchenmitglieder, Katholiken und Bevölkerung nach Bundesländern am 31 December 2007. Retrieved 14 April 2010.

[94] Die Deutsche Bischofskonferenz – German Catholic Church, Population and Catholics by Bundesstaat (http://www.zenit.org/article-28425?l=english). Retrieved 14 April 2010.

[95] de:Jüdisches Leben in Berlin

[96] Germany: Berlin Facing Challenge Of Assimilating Russian-Speaking Jews (http://www.rferl.org/content/Article/1078688.html). Radio Free Europe. 17 September 2007.

[97] "Lutheran Diocese Berlin-Brandenburg" (http://www.selk-berlin.de/). *Selbständige Evangelisch-Lutherische Kirche*. . Retrieved 19 August 2008.

[98] "Jahrgangsstufe Null" (http://www.tagesspiegel.de/berlin/;art270,2185300) (in German). *Der Tagesspiegel*. . Retrieved 19 August 2008.

[99] "Geschichte des Französischen Gymnasiums" (http://www.fg-berlin.de/WebObjects/FranzGym.woa/wa/CMSshow/1064384) (in German). *Französisches Gymnasium Lycée Français Berlin*. . Retrieved 17 August 2008.

[100] "Metropolis of Sciences" (http://web.archive.org/web/20080424003113/http://www.berlin-partner.de/622/?L=1). *Berlin Partner GmbH*. Archived from the original (http://www.berlin-partner.de/622/?L=1) on 24 April 2008. . Retrieved 19 August 2008.

[101] "Wirtschaftsstandort Berlin" (http://www.tagesspiegel.de/berlin/abgeschlagen-auf-den-hinteren-raengen/1820474.html) (in German). *Der Tagesspiegel*. 11 May 2010. . Retrieved 11 May 2010.

[102] "European Institute of Innovation and Technology: Home" (http://eit.europa.eu/home.html). *Europa (web portal)*. . Retrieved 8 June 2010.

[103] "World Heritage Site Palaces and Parks of Potsdam and Berlin" (http://whc.unesco.org/en/list/532). *UNESCO*. . Retrieved 19 August 2008.

[104] "Hub Culture's 2009 Zeitgeist Ranking" (http://www.hubculture.com/groups/hubnews/news/162/). *Hub Culture*. . Retrieved 30 April 2009.

[105] "Sprung in die Wolken" (http://service.zitty.de/kultur-kunst/14539/) (in German). *Zitty*. 2 July 2008. . Retrieved 19 August 2008.

[106] Boston, Nicholas (10 September 2006). "A New Williamsburg! Berlin's Expats Go Bezirk" (http://www.observer.com/node/39370). *The New York Observer*. . Retrieved 17 August 2008. See also: "Die Kunstszene" (http://web.archive.org/web/20071211095052/http://www.magazine-deutschland.de/magazin/J-Kunstszene_2-05.php) (in German). *Deutschland Online*. Archived from the original (http://www.magazine-deutschland.de/magazin/J-Kunstszene_2-05.php) on 11 December 2007. . Retrieved 19 August 2008. and "Culture of Berlin" (http://web.archive.org/web/20070928020742/http://www.metropolis2005.org/en/berlin/kultur.html). *Metropolis*. Archived from the original (http://www.metropolis2005.org/en/berlin/kultur.html) on 28 September 2007. . Retrieved 19 August 2008.

[107] "Saucy Berlin transforms itself into a 'music city'" (http://www.taipeitimes.com/News/feat/archives/2005/09/05/2003270493). *Taipei Times*. . Retrieved 19 August 2008.

[108] "Berlin's music business booms" (http://web.archive.org/web/20070911125347/http://www.stadtentwicklung.berlin.de/bauen/bueroflaechen/en/friedrichshain.shtml). *Expatica*. Archived from the original (http://www.stadtentwicklung.berlin.de/bauen/bueroflaechen/en/friedrichshain.shtml) on 11 September 2007. . Retrieved 19 August 2008.

[109] "Media Companies in Berlin and Potsdam" (http://www.medienboard.de/WebObjects/Medienboard.woa/wa/CMSshow/2809830). *medienboard*. . Retrieved 19 August 2008.

[110] "Wall-to-wall culture" (http://www.theage.com.au/news/arts/walltowall-culture/2007/11/09/1194329483873.html). *The Age* (Australia). 10 November 2007. . Retrieved 30 November 2007.

[111] European Film Academy (http://www.europeanfilmacademy.org/), www.europeanfilmacademy.org, Accessed 19 December 2006. See also: Berlin Film Festival (http://www.berlinale.de/), www.berlinale.de. Retrieved 12 November 2006.

[112] Wasacz, Walter (11 October 2004). "Losing your mind in Berlin" (http://www.metrotimes.com/editorial/story.asp?id=6949). *Metro Times*. . Retrieved 18 November 2006.

[113] Compensating Victims of the La Belle Attack (http://thomaspainescorner.org/?p=373), German Embassy, Washington D.C.. Retrieved 18 November 2006.

[114] Christiane F.-Page (http://christiane.hurz.at/), christianef. Retrieved 18 November 2006.
[115] ENGLISH SUMMARY (http://www.karneval-berlin.de/de/english.175.html), www.karneval-berlin.de, . Retrieved 10 August 2008.
[116] Berlin for Gays and Lesbians (http://web.archive.org/web/20061007101504/http://www.berlin-tourist-information.de/english/zielgruppen/e_zg_gay_bezirke.php), Berlin Tourismus Marketing GmbH. Retrieved 20 October 2006.
[117] Krauss, Kenneth (2004). *The drama of fallen France: reading la comédie sans tickets*. Albany: State University of New York. p. 11. ISBN 0-7914-5953-5.
[118] "Berlin Lesbian and Gay Festival 2011 – Berlin Gay Pride 2011" (http://gaytravel.about.com/od/previewsofpridefestivals/qt/Berlin_Gay_Pride.htm). 25 July 2008. .
[119] "THE GAY CAPITAL OF EUROPE BERLIN" (http://www.passportmagazine.com/destinations/Berlin769.php). 25 September 2010. .
[120] A 3,000-year-old smile (http://www.expatica.com/de/life_in/leisure/a-3000-year-old-smile-the-strange-history-of-queen-nefertiti-22752.html), Expatica.Com. Retrieved 1 November 2006.
[121] "Exhibitions" (http://web.archive.org/web/20071011005147/http://www.juedisches-museum-berlin.de/site/EN/01-Exhibitions/exhibitions.php). *Jewish Museum Berlin*. Archived from the original (http://www.juedisches-museum-berlin.de/site/EN/01-Exhibitions/exhibitions.php) on 11 October 2007. . Retrieved 10 August 2008.
[122] The World of Dinosaurs (http://www.naturkundemuseum-berlin.de/en/ausstellungen/the-world-of-dinosaurs.html?Fsize=0&Lightversion=0?typo=2?typo=1?typo=0), Museum für Naturkunde. Retrieved 10 August 2008.
[123] In Berlin, the Art of Sex (http://www.washingtonpost.com/wp-srv/travel/index/stories/beck04181999.htm), Washington Post. Retrieved 10 August 2008.
[124] Is Rattle's Berlin honeymoon over? (http://www.guardian.co.uk/world/2004/apr/29/germany.arts), The Guardian. Retrieved 12 November 2006.
[125] Wakin, Daniel J. (25 September 2005). "Music: Berlin" (http://travel2.nytimes.com/2005/09/25/travel/25berlin.html?ex=1162962000&en=36217d7f20459f88&ei=5070). *The New York Times*. . Retrieved 7 November 2006.
[126] Berlin Philharmonic elects Sir Simon Rattle (http://www.culturekiosque.com/klassik/news/rherattle.htm), Culturekiosque. Retrieved 12 November 2006.
[127] Haus der Kulturen der Welt (http://www.hkw.de/en/bottom/impressum/impressum.php), www.hkw.de. Retrieved 12 November 2006.
[128] Berlin Minimalist Glamor (http://www.nytimes.com/2010/11/09/fashion/09iht-luxmag18.html) Retrieved
[129] Paterson, Tony (15 August 2009). "Spicy sausage that is worthy of a shrine in Berlin" (http://www.independent.co.uk/life-style/food-and-drink/news/spicy-sausage-that-is-worthy-of-a-shrine-in-berlin-1772530.html). *The Independent* (London). .
[130] "Berlin Style Liver – Gebratene Kalbsleber auf Berliner Art, Germanic" (http://www.germancorner.com/recipes/A-B/berlin_style_liver_.html). *Germancorner.com*. . Retrieved 8 June 2010.
[131] Berlin (http://www.germanfoods.org/consumer/facts/berlin.cfm) German Foods
[132] "Hauptstadt-Zoo beliebtester Tierpark" (http://web.archive.org/web/20061007155448/http://www.rbb-online.de/_/nachrichten/vermischtes/beitrag_jsp/key=news4382800.html). *Rundfunk Berlin-Brandenburg*. Archived from the original (http://www.rbb-online.de/_/nachrichten/vermischtes/beitrag_jsp/key=news4382800.html) on 7 October 2006. . Retrieved 17 August 2008.
[133] Moore, Tristana (23 March 2007). "Baby bear becomes media star" (http://news.bbc.co.uk/1/hi/world/europe/6486993.stm). *BBC News*. . Retrieved 17 August 2008.
[134] Boyes, Roger (13 December 2007). "Berlin Zoo culls creator of the cult of Knut" (http://www.timesonline.co.uk/tol/news/world/europe/article3042791.ece). *The Times* (UK). . Retrieved 17 August 2008.
[135] "Grün Berlin [Green Berlin]" (http://www.gruen-berlin.de/gesellschaft/) (in German). Die Grün Berlin GmbH. . Retrieved 27 May 2011.
[136] Peter Joseph Lenné (http://www.stadtentwicklung.berlin.de/aktuell/wettbewerbe/lenne/en/biographie.shtml), Senate Department of Urban Development. Retrieved 18 November 2006.
[137] 36 Hours in Berlin (http://travel.nytimes.com/2006/12/10/travel/10hours.html), NYT, Accessed 29 May 2007.
[138] "Melbourne retains ultimate sports city title" (http://www.abc.net.au/news/stories/2008/04/01/2204562.htm?section=sport). *ABC News*. 1 April 2008. . Retrieved 1 July 2008.
[139] Berlin 1936 Games of the XI Olympiad (http://www.olympic.org/uk/games/past/index_uk.asp?OLGT=1&OLGY=1936), www.olympic.org, Accessed 18 November 2006. See also: "Italy conquer the world as Germany wins friends" (http://www.fifa.com/worldcup/archive/germany2006/overview.html). .
[140] 12. IAAF Leichtathletik WM berlin 2009 (http://www.berlin2009.org/), www.berlin2009.org. Retrieved 1 July 2008.
[141] Berlin Marathon (http://www.scc-events.com/en/events/halbmarathon/), www.scc-events.com. Retrieved 12 November 2006.
[142] 500,000 spectators to watch the game together (http://blogs.bettor.com/The-best-sports-stadiums-in-the-world-a21645), SABAH Newspaper. Retrieved 1 July 2008.
[143] Hertha BSC (http://www.herthabsc.de/), www.herthabsc.de/. Retrieved 1 July 2008.
[144] Union Berlin (http://www.fc-union-berlin.de/), www.fc-union-berlin.de/. Retrieved 3 April 2010.
[145] ALBA Berlin (http://www.albaberlin.de/), www.albaberlin.de. Retrieved 1 July 2008.
[146] Eisbären Berlin (http://www.eisbaeren.de/content/index.php), www.eisbaeren.de. Retrieved 1 July 2008.
[147] Füchse Berlin (http://www.fuechse-berlin.de/), www.fuechse-berlin.de. Retrieved 1 July 2008.
[148] SCC Berlin (http://www.scc-berlin.de/abteilungen.html), www.scc-berlin.de. Retrieved 1 July 2008.

[149] (http://www.volleyball-bundesliga.de/volleyball/ligen.php?menuid=192&topmenu=395&liga=1&runden=130&keepmenu=inactive) Retrieved 21 May 2010

[150] Scheffler, Karl. *Berlin: Ein Stadtschicksal*. Fannei & Walz. ISBN 3927574023.

[151] "Citysongs" (http://www.newcolonist.com/cs-berlin.html). *The New Colonist*. . Retrieved 18 August 2008.

[152] "David Bowie's golden years in Berlin" (http://www.expatica.com/de/life_in/feature/david-bowies-golden-years-in-berlin-31468.html). *Expatica*. . Retrieved 18 August 2008.

[153] Tomforde, Anna; Farr, Michael (11 November 1999). "Kohl: We are one nation" (http://www.guardian.co.uk/world/1989/nov/11/berlinwall.germany3). *The Guardian* (UK). . Retrieved 18 August 2008.

[154] "Arm, aber sexy" (http://www.focus.de/politik/deutschland/wowereits-berlin-slogan_aid_117712.html) (in German). *Focus*. . Retrieved 18 August 2008.

References

- Chandler, Tertius (1987). *Four Thousand Years of Urban Growth: An Historical Census*. Edwin Mellen Pr. ISBN 0889462070.
- Gill, Anton (1993). *A Dance Between Flames: Berlin Between the Wars*. John Murray. ISBN 0719549868.
- Gross, Leonard (1999). *The Last Jews in Berlin*. Carroll & Graf Publishers. ISBN 0786706872.
- Large, David Clay (2001). *Berlin*. Basic Books. ISBN 046502632X.
- Read, Anthony; David Fisher (1994). *Berlin Rising: Biography of a City*. W.W. Norton. ISBN 0393036065.
- Ribbe, Wolfgang (2002). *Geschichte Berlins*. Bwv – Berliner Wissenschafts-Verlag. ISBN 3830501668.
- Roth, Joseph (2004). *What I Saw: Reports from Berlin 1920–33*. Granta Books. ISBN 1-86207-636-7.
- Taylor, Frederick (2007). *The Berlin Wall: 13 August 1961 – 9 November 1989*. Bloomsbury. ISBN 0060786140.

External links

- Berlin.de (http://www.berlin.de/international/index.en.php/)—Official Website
- Official Website of the Berlin-Brandenburg Metropolitan Area (http://www.hauptstadtregion-berlin-brandenburg.de/)
- Regional directory (http://www.berlin-regional.de/) (German)
- Berlin Tourist Information (http://www.berlinpass.com/)
- Dumont Restaurantguide Berlin (http://www.gastro-stars-berlin.com/)
- Monthly English-language magazine for Berlin (http://www.exberliner.com/)
- English-language city guide for Berlin (http://berlin.unlike.net/)
- Berlin Travel Tips (http://iwannagothere.com/travel-guides/berlin/11)
- Berlin travel guide from Wikitravel
- Berlin photo gallery (http://www.flickr.com/photos/41182695@N08/sets/72157622725699696/)—Images of Berlin and its most popular monuments.

kbd:Берлин gag:Berlin mrj:Берлин ltg:Berlins xmf:ბერლინი pfl:Balin koi:Берлин rue:Берлін

Berlin S-Bahn

S-Bahn Berlin

Berlin Friedrichstrasse railway station, crossing point for the Stadtbahn and the Nord-Süd-Tunnel routes of the Berlin S-Bahn

Info	
Locale	Berlin
Transit type	Rapid transit
Number of lines	15
Number of stations	166
Daily ridership	1,060,000
Operation	
Began operation	1924
Operator(s)	S-Bahn Berlin GmbH
Technical	
Track gauge	1435 mm (4 ft 8 $\frac{1}{2}$ in) (standard gauge)

The **Berlin S-Bahn** ("Schnellbahn", literally "fast train") is a rapid transit system in and around Berlin, the capital city of Germany. It consists of 15 lines and is integrated with the mostly underground U-Bahn to form the backbone of Berlin's rapid transport system. Unlike the U-Bahn, the S-Bahn crosses the Berlin city and state border into the surrounding state of Brandenburg, mostly from the former East Berlin but today also from West Berlin to Potsdam.

The Berlin S-Bahn network

Although the S- and U-Bahn are part of a unified fare system, they have different operators. The S-Bahn is operated by *S-Bahn Berlin GmbH*, a subsidiary of the Deutsche Bahn, whilst the U-Bahn is run by BVG, the main public transit company for the city of Berlin.

Operation

Routes

The S-Bahn routes all feed into one of three core lines: a central, elevated east-west line (*the Stadtbahn*), a central, mostly underground north-south line (*the Nord-Süd-Tunnel*), and a circular, elevated line (*the Ringbahn*). Geographically, the Ringbahn takes the form of a dog's head and is colloquially known to Berliners by that name (*Hundekopf*). Outside the Ringbahn, suburban routes radiate out in all directions.

Generally speaking, the first digit of a route number designates the main route or a group of routes. Thus, S25 is a bifurcation of S2, while S41, S42, S45, S46, and S47 are all Ringbahn routes that share some of the same route.

Line	Terminus	Route	Terminus
S1	Wannsee	**Nord-Süd-Tunnel**	Oranienburg
S2	Blankenfelde	**Nord-Süd-Tunnel**	Bernau
S25	Teltow Stadt	**Nord-Süd-Tunnel**	Hennigsdorf
S3	Erkner	**Stadtbahn**	Spandau
S41	Südkreuz	**Ringbahn**	Südkreuz (clockwise)
S42	Südkreuz	**Ringbahn**	Südkreuz (counter-clockwise)
S45	Berlin-Schönefeld	**Ringbahn**	Südkreuz (↔ Bundesplatz)
S46	Königs Wusterhausen	**Ringbahn**	Westend
S47	Spindlersfeld	**Ringbahn**	Hermannstraße (↔ Südkreuz)
S5	Strausberg Nord	**Stadtbahn**	Westkreuz
S7	Ahrensfelde	**Stadtbahn**	Potsdam Hauptbahnhof
S75	Wartenberg	**Stadtbahn**	Spandau
S8	(Zeuthen ↔) Grünau	**Ringbahn**	Hohen Neuendorf
S85	(Grünau ↔) Schöneweide	**Ringbahn**	Waidmannslust
S9	Berlin-Schönefeld	**Ringbahn**	Blankenburg

Stations in brackets are serviced at certain times only (Monday-Friday during offpeak in the case of S47 and during peak in the case of S8 and S85). S45 and S85 only run Mon-Fri.

Also, not every train reaches the nominal terminus of a line. For example, every other train on S1 runs only to Frohnau, five stops before Oranienburg, and the last stop on S3 towards Erkner which is reached by every train is

Friedrichshagen. Similarly, some of the S2 trains terminate northwards only at Gesundbrunnen, and most of S5 trains run only to Strausberg or even Mahlsdorf, rendering Strausberg Nord the least frequented stop on the whole network.

On 31 August 2009 a few permanent changes to the line routes were applied. Due to renovation of the Ostkreuz station, which includes dismantling tracks connecting the Stadtbahn and the Ringbahn, S9 (formerly Berlin-Schönefeld ↔ Spandau) cannot turn west at this station any more. The line thus follows the Ringbahn and then branches northwards past Schönhauser Allee, like S2 and S8, and terminates at Blankenburg. To compensate for the diminished throughput on the Stadtbahn, the S3 (formerly Erkner ↔ Ostbahnhof) is extended westwards to Spandau. Minor alterations have been applied to the routes of S45 and S47, and S85 has been suspended on weekends.

Service hours

The normal daytime service runs fundamentally between 04:00 and 01:00 Monday-Friday, between 05:00 and 01:00 on Saturdays and between 06:30 and 01:00 on Sundays. However, there is a comprehensive nighttime service on most lines between 01:00 and 05:00 on Saturdays and 01:00 and 06:30 on Sundays, which means that most stations enjoy a continuous service between Friday morning and Sunday evening. One exception to this is the section of the S8 between Blankenburg and Hohen Neuendorf which has no service during these times. Many other lines are totally unchanged in their operation, but some are curtailed or extended during nighttime service. Particularly, the S1, S2, S25, S41, S42, S75 are unchanged, and the S45, S85 have no nighttime service. The only line that is extended rather than curtailed is S5, which then runs westward as far as Potsdam Hbf to compensate for S7 servicing only the Ahrensfelde ↔ Lichtenberg section. Remaining lines S3, S46, S47 and S9 terminate westwards at stations Ostkreuz, Südkreuz, Schöneweide and Treptower Park (northwards), respectively.

History

Inception

With individual sections dating from the 1870s, the S-Bahn came into existence in 1924. It was formed as the network of suburban commuter railways running into Berlin was converted from steam operation to a third-rail electric railway. The resulting network was primarily above-ground but with some subsurface tunnels.

Services on the Berlin S-Bahn were at first provided by the German national railway, the Deutsche Reichsbahn. Electrification of the existing suburban lines was completed around 1929, and thoughts turned to a new project: a tunnel that would join two spur lines that protruded into the city centre from the north and south. This tunnel, to be known as the Nord Süd Bahn, was a prestige project for the Nazis, and was opened in two sections. The first, from the north to Unter den Linden, opened in time for the 1936 Berlin Olympics; the final section, via Potsdamer Platz, opened the month after the Second World War began, in October 1939.

During and after World War II

Many sections of the S-Bahn were closed during the war due to enemy action. The Nord Süd Bahn tunnel was flooded on 2 May 1945 by retreating SS troops during the final Battle of Berlin. The exact number of casualties is not known, but up to 200 persons are presumed to have perished, since the tunnel was used as a public shelter and also served to house military wounded in trains in underground sidings. Service through the tunnel commenced again in 1947.

Some Type 477 trains, built before World War II, remained in service until the early 21st century

After hostilities ceased in 1945, Berlin was given special status as a "Four Sector City," surrounded by the Soviet Occupation Zone, which later became the German Democratic Republic (GDR). The Allies had decided that S-Bahn service in the western sectors of Berlin should continue to be provided by the Reichsbahn (DR), which was by now the provider of railway services in East Germany. (Rail services in West Germany proper were provided by the new Deutsche Bundesbahn.)

During the war, Berlin S-Bahn cars were overhauled at Luben to the east of Berlin. As that town, now known as Lubin, was ceded to Poland under the terms of the Potsdam Conference in 1945, 84 cars currently in the works were lost to Berlin. Further cars were sent east as war reparations, and eventually at least 287 cars were sent to Russia where they were converted for use in Moscow, Kiev and Tallinn. Additionally at least 80 two-car sets were retained in Poland, where they were used on suburban services in the Gdańsk-Gdynia region until 1976. Some of the latter cars were then converted for use in overhead line maintenance trains, and some still exist in that role. One set is preserved in its Gdańsk-Gdynia condition at a museum at Koscierzyna near Gdynia.[1]

Cold War

As relations between East and West began to sour with the coming of the Cold War, the Berlin S-Bahn soon became a victim of the hostilities. Although services continued operating through all occupation sectors, checkpoints were constructed on the borders to East Berlin and on-board "customs checks" were carried out on trains. From 1958 onwards, some S-Bahn trains ran non-stop through the western sectors from stations in East Berlin to stations on outlying sections in East Germany, so as to avoid the need for such controls. From this point onwards, all East German government employees were forbidden to use the S-Bahn as it travelled through West Berlin.

Alexanderplatz is an important transport hub in eastern Berlin

The western sectors of the city were physically cut off from East Germany on 13 August 1961 by what was later called the Berlin Wall, in a well-prepared plan to separate the two halves of the city – and at the same time, to divide the Berlin public transit network into two separate systems. Stadtbahn services were curtailed from both directions at the Friedrichstraße station. This station was divided into two physically separated areas, one for eastern passengers and one for westerners. Although the station lay within East Berlin, western passengers could transfer between S-Bahn lines or to the U-Bahn without passing through border checks, much like passengers changing planes at an international airport. The GDR also operated an Intershop in the portion of the station with services to and from West Berlin , where persons arriving from West Berlin (again without passing through border controls) could buy luxury goods such as tobacco and alcoholic beverages at discounted prices (compared to prices in West Berlin), provided they paid in hard currency, owing in part to the fact that Intershop customers did not pay West German taxes on their purchases. The West Berlin authorities were aware of this situation but did not impose stringent customs controls on such purchases out of political considerations. The Friedrichstraße station also become the main entry point for train and subway riders from West Berlin into East Berlin. Service on the Nord Süd Bahn was operated for western passengers only. It passed through a relatively short stretch under East Berlin territory in the city centre, and trains did not stop at the underground East Berlin S-Bahn stations, which were called ghost stations. Similarly, the Ringbahn services in the north of the city were curtailed at Gesundbrunnen from the west and Schönhauser Allee from the east, and in the south-east of the city at Sonnenallee and Köllnische Heide from the west and Treptower Park and Baumschulenweg from the East.

Because the S-Bahn was operated by the DR, West Berliners vented their frustration at the building of the wall by boycotting it since its fares were seen as subsidising the communist regime in the East. "Keinen Pfennig mehr für Ulbricht," or "not a penny more for Ulbricht," became the S-Bahn opponents' chant. Within days of the Berlin Wall being built, the BVG, with assistance from other transit companies in West Germany, began providing "solidarity with Berlin buses" – new bus services which paralleled the S-Bahn lines and therefore provided an alternative. After many years of declining passenger usage and difficult industrial relations between the West Berlin workforce and their East Berlin employers, most of the western portion of the S-Bahn was closed down in September 1980 following a strike. A 20-minute service was still provided on the Stadtbahn from Westkreuz to Friedrichstraße as well as services on the Nord-Süd Bahn between Frohnau, Friedrichstraße, Lichtenrade or Wannsee.

By contrast, during the same period, services on the S-Bahn in East Berlin were increased and new lines built as housing projects expanded eastward from the city centre. With most of the U-Bahn located in West Berlin, the S-Bahn became the backbone of the East Berlin transit network.

The track system is a fully separated and independent rapid transport system within the city. Station: Anhalter Bahnhof

The 1980 incidents turned media and political attention towards what was left of West Berlin's S-Bahn network. The city government decided to enter negotiations with East Germany which were finally successful. On 9 January 1984, the BVG took over the responsibility for operation of S-Bahn services in West Berlin. After further closedowns that same day, a limited service was restored, initially comprising only two short sections without direct interchange between them. In the years between 1984 and 1989, several sections were gradually reopened, resulting in a network of 71 km and three lines - with one line running on the Stadtbahn and two on the Nord Süd Bahn - comprising about 50% of West Berlin's original network. This development brought West Berlin's S-Bahn back into public awareness and restored its popularity.

Until 1984, all Berlin S-Bahn routes were allocated letters as a means of identifying the route of the train. These letters were occasionally followed by Roman numerals to indicate a shortworking or bifurcation in the service (e.g., A, BI, BII, C,) and are still used internally by the Berlin S-Bahn GmbH for timetabling and in conjunction with radio call-signs to each train unit. When the BVG took over the responsibility for operation of S-Bahn services in West Berlin in 1984, it introduced a new unified numbering scheme for both the S-Bahn and the U-Bahn, which it also operated. Existing U-Bahn route numbers were prefixed with the letter *U,* while the new S-Bahn route numbers were prefixed with the letter *S.* This system of numbering routes was used in other West German cities, and was extended to the S-Bahn service for the whole city after reunification.

Reunification

A modern S-Bahn train at Griebnitzsee

After the Berlin Wall came down in November 1989, the first broken links were re-established, with Friedrichstraße on July 1, 1990 as the first. The BVG and DR jointly marketed the services soon after *die Wende*. Administratively, the divided S-Bahn networks remained separate in this time of momentous changes, encompassing German reunification and reunification of Berlin into a single city, although the dividing line was no longer the former Berlin Wall. DR and BVG (of the whole of reunified Berlin from January 1, 1992 after absorbing BVB of East Berlin) operated individual lines end to end, both into the other party's territories. For example, S2 was all-BVG even after it was extended northward and southward into Brandenburg/former GDR territory. The main east-west route (Stadtbahn) was a joint operation. Individual trains were operated by either BVG or DR end-to-end on the same tracks. This arrangement ended on January 1, 1994 with the creation of Deutsche Bahn due to the merger between DR and the former West Germany's Deutsche Bundesbahn. All S-Bahn operations in Berlin were transferred to the newly formed S-Bahn Berlin GmbH as a subidiary of Deutsche Bahn and the BVG withdrew from running S-Bahn services.

Technically, a number of projects followed in the steps of re-establishing broken links in order to restore the former S-Bahn network to its 1961 status after 1990, especially the Ringbahn. In December 1997 the connection between Neukölln and Treptower Park via Sonnenallee was reopened, enabling S4 trains to run 75% of the whole ring between Schönhauser Allee and Jungfernheide. On 16 June 2002, the section Gesundbrunnen - Westhafen also reopened, re-establishing the Ringbahn operations.

Service reductions

On 20 July 2009, known locally as "Black Monday," the S-Bahn service was significantly reduced due to safety checks on the trains ordered by the German Federal Railway Authority. Having so many trains taken out of service for inspection left less than 30 percent of the system's rolling stock available for revenue service. Eight routes, including most through services on the Stadtbahn, were closed and on other lines headways were reduced to 20 minutes and trains shortened.[2]

Some minor restorations in service were made on 3 August 2009. Due to new inspection troubles the S-Bahn network was again reduced dramatically on 8 September 2009. As of that date, three quarters of the trains were withdrawn from the network due to inspection and faulty brake cylinders.[3] There were again no trains on the Stadtbahn between Westkreuz and Alexanderplatz, and no S-Bahn trains to Spandau. Trains on the circle lines, S41 and S42, were running at 10-minute intervals. Other routes were running with extended intervals and reduced distances.[4]

In late 2009, the Berlin Senate expected that normal operations would only resume in 2013.[5] In January 2010, DB announced that they expected the system to resume normal service in December 2010 and employed 300 new staff in their workshops.[6] In the same month, the Berlin transport Senator Ingeborg Junge-Reyer rejected an extension of the traffic contract with the operator Deutsche Bahn (DB) which is due to expire in December 2017. Different routes for the operation of the S-Bahn from 2018 are being examined. These include gradual tender lines, splitting the system up into subnetworks or the acquisition of the S-Bahn by the Berlin Lander.[7]

By Spring of 2011, some 420 train sets were in service, a considerable improvement over the situation in 2009, but still insufficient compared to the 500 needed to provide a normal full service. The S-Bahn announced it was to invest 120 million euros in order to achieve 500 train sets in service by December of 2011.[8] Rüdiger Grube, the head of the DB, announced that losses due to the S-Bahn crisis had reached 370 million euros at the end of 2010. He expected them to reach 700 million euros by the end of 2014, with no operating profits to be made before the end of the contract in December 2017.[9]

Infrastructure work

Starting in 2010, DB Netz is replacing mechanical train stops on the S-Bahn network with electronic balises.[10]

See also

- List of Berlin S-Bahn stations
- List of metro systems

References

[1] Fender, Keith; Bent, Mike (February 2011). "Old Berlin/Gdansk S-Bahn cars in museum and in use". *Today's Railways* (Platform 5 Publishing Ltd): pp. 61.
[2] "Chaos in Berlin. Eisenbahn-Bundesamt legt S-Bahnen still" (http://www.spiegel.de/reise/aktuell/0,1518,633763,00.html). Spiegel Online. 1 July 2009. . Retrieved 1 May 2011.
[3] "Berliner S-Bahn droht neues Chaos" (http://www.spiegel.de/reise/deutschland/0,1518,647539,00.html). Spiegel Online. . Retrieved 2011-05-01.
[4] "Wie die S-Bahn in Schieflage geriet [As the train ran into difficulties]" (http://www.tagesspiegel.de/berlin/S-Bahn-Deutsche-Bahn;art270,2974104) (in German). Der Tagesspiegel. 2009-12-12. . Retrieved 2011-05-01.
[5] "S-Bahn fährt frühestens 2013 wieder normal [S-Bahn not back to normal until at least 2013]" (http://www.tagesspiegel.de/berlin/Verkehr-S-Bahn-Deutsche-Bahn;art18614,2986470) (in German). Der Tagesspiegel. 2009-12-28. . Retrieved 2011-05-01.
[6] "Deutsche Bahn: 70 Mio. Euro zusätzliche Entschädigung für S-Bahn-Kunden – Normalisierter Betrieb bis Ende 2010 [Deutsche Bahn: € 70 million additional compensation for S-Bahn customer - Normalized operating by end 2010]" (http://www.s-bahn-berlin.de/presse/presse_anzeige.php?ID=535) (in German). S-Bahn Berlin. .
[7] "Berlin S-Bahn-Chaos "Ich glaube der Bahn nichts mehr"" (http://www.sueddeutsche.de/wirtschaft/192/499469/text). Süddeutsche Zeitung. 2010-12-07. .
[8] "Worldwide Review - Germany - Berlin". *Tramways & Urban Transit* (Ian Allan Ltd / Light Rail Transit Association): pp. 193. May 2011.
[9] "Berliner S-Bahn-Chaos kostet 700 Millionen Euro [Berlin S-Bahn chaos will cost 700 million euros]" (http://www.spiegel.de/wirtschaft/service/0,1518,738665,00.html) (in German). Spiegel-Online. 2010-01-10. .
[10] "Railway Gazette: Berlin S-Bahn control update" (http://www.railwaygazette.com/news/single-view/view/10/berlin-s-bahn-control-update.html). . Retrieved 2010-08-18.

External links

- S-Bahn Berlin GmbH (http://www.s-bahn-berlin.de/englisch/home.html)
- regular network (PDF) (http://www.s-bahn-berlin.de/pdf/s_bahn_netz.pdf)
- VSWB.de - Disused Rails and Lanes in Berlin (http://berlin.bahninfo.de/E_index.htm)
- Google Mashup with S-Bahn Map (beta) (http://www.uberbahn.com/)
- Flyer containing special map due to the cut traffic situation since July 20, 2009 (http://vbb.kontur-bb.de/flyer.pdf)

Commuter rail

Commuter rail, also called **suburban rail**, is a passenger rail transport service that primarily operates between a city center, and the middle to outer suburbs beyond 15km (10 miles) and commuter towns or other locations that draw large numbers of commuters—people who travel on a daily basis. Trains operate following a schedule, at speeds varying from 50 to 200 km/h (30 to 125 mph). Distance charges or zone pricing may be used.

The Long Island Rail Road operates electric and diesel service into New York City.

Non-English names include *Treno suburbano* in Italian, *Cercanías* in Spanish, *Nahverkehrszug* in German (and in most larger cities *S-Bahns* though these trains also often include city centre metro-like sections where lines have merged together and services become more frequent, and stations are closer together to better distribute passengers into the city core),*Train de banlieue* in French, *Příměstský vlak* in Czech and Elektrichka in Russian. The development of commuter rail services has become popular today, with the increased public awareness of congestion, dependence on fossil fuels, and other environmental issues, as well as the rising costs of owning, operating and parking automobiles.

Characteristics

İZBAN operates commuter rail service in İzmir, Turkey.

Most commuter (or suburban) trains are built to main line rail standards, differing from light rail or rapid transit (metro rail) systems by:

- being larger
- providing more seating and less standing room, for the longer distances involved
- having (in most cases) a lower frequency of service
- having scheduled services (i.e. trains run at specific times rather than at specific intervals)
- serving lower-density suburban areas, typically connecting suburbs to the city center
- sharing track or right-of-way with intercity or freight trains
- not fully grade separated (containing at-grade crossings with crossing gates)

Train schedule

A First ScotRail operated Class 318 departing Gourock.

Compared to rapid transit (or metro rail), commuter/suburban rail has lower frequency, following a schedule rather than fixed intervals, and fewer stations spaced further apart. They primarily serve lower density suburban areas (non inner-city), and often share right-of-way with intercity or freight trains. Some services operate only during peak hours. Average speeds are high, often 50 km/h (30 mph) or higher. These higher speeds better serve the longer distances involved. Some services include express services which skip some stations in order to run faster and separate longer distance riders from short-distance ones.

A Cercanias train near Cádiz, Spain.

The general range of commuter trains' distance varies between 15 and 200 km (10 and 125 miles).

Track

A Southeastern commuter train at Slade Green in South East London, England, running a service to London Cannon Street.

Their ability to coexist with freight or intercity services in the same right-of-way can drastically reduce system construction costs. However, frequently they are built with dedicated tracks within that right-of-way to prevent delays, especially where service densities have converged in the inner parts of the network.

Most such trains run on the local standard gauge track. Some light rail systems may run on a narrower gauge. Examples of narrow gauge systems are found in Japan, Switzerland, in the Brisbane (Citytrain) and Perth (Transperth) systems in Australia, and on the Genoa-Casella line in Italy. Some countries, including India, Pakistan, Russia, Brazil and Sri Lanka, as well as Melbourne and Adelaide in Australia, use broad gauge.

Distinction between other modes of rail

South Korean Commuter Car, Tonggeun, in Dorasan Station.

Metro

Metro rail or rapid transit usually covers a smaller inner-urban area ranging outwards to between 12km to 20km (or 8 to 14 miles), has a higher train frequency and runs on separate tracks (underground or elevated), whereas commuter rail often shares tracks, technology and the legal framework with mainline railway.

However, the classification as a metro or rapid rail can be difficult as both may typically cover a metropolitan area exclusively, run on separate tracks in the centre, and often feature purpose-built rolling stock. The fact that the terminology is not standardised across countries (even across English-speaking countries) further complicates matters. The distinction is most easily made when there are two systems such as New York's subway and the LIRR and Metro-North, Paris' RER and Métro, London's Overground and the tube lines of the Underground, the S-Bahn and U-Bahn systems in Berlin, Frankfurt, Munich or Vienna, the JR lines and the Metro in Tokyo.

Two electric multiple units of the class 423 common in several German S-Bahn networks.

Regional rail

Regional rail usually provides rail services between towns and cities, rather than purely linking major population hubs in the way inter-city rail does. Regional rail operates outside major cities. Unlike Inter-city, it stops at most or all stations between cities. It provides a service between smaller communities along the line, and also connections with long-distance services at interchange stations located at junctions or at larger towns along the line. Alternative names are "local train" or "stopping train". Examples include the former BR's Regional Railways, France's TER (*Transport express régional*) and Germany's DB Regio services. Regional rail does not exist in this sense in the United States, so the term "regional rail" has become synonymous with commuter rail there, although the two are more clearly defined in Europe.

Cab control car of a double-deckered GO Train in Toronto.

Inter-city rail

In some European countries the distinction between commuter trains and long-distance/intercity trains is very hard to make, because of the relatively short distances involved. For example, so called "intercity" trains in Belgium and the Netherlands carry many commuters and their equipment, range and speeds are similar to those of commuter trains in some larger countries.

Russian commuter trains, on the other hand, frequently cover areas larger than Belgium itself, although these are still short distances by Russian standards. They have a different ticketing system from long-distance trains, and in major cities they often operate from a separate section of the train station.

Elektrichka departing from a station platform in Russia

The easiest way to identify these "Inter-city" services is that they tend to operate as express services - only linking the main stations in the cities they link, not stopping at any other stations. However, this term is used in Australia (Sydney for example) to describe the regional trains operating beyond the boundaries of the suburban services, even though some of these "inter-city" services stop all stations similar to German regional services. In this regard, the German service delineations and corresponding naming conventions are clearer and better used for academic purposes.

High-speed rail

Mumbai Suburban Railway, the lifeline of Mumbai, carries more than 6.9 million commuters on a daily basis

Sometimes high-speed rail can serve daily use of commuters. The Japanese Shinkansen high speed rail system is heavily used by commuters in the Greater Tokyo Area. They commute between 100 and 200 km by Shinkansen. To meet the demand of commuters, JR sells commuter discount passes and operates 16 car bilevel E4 Series Shinkansen at rush hour, providing a capacity of 1,600 seats. Several lines in China such as Guangzhou-Zhuhai Intercity Mass Rapid Transit, the Beijing–Tianjin Intercity Railway, and the Shanghai–Nanjing High-Speed Railway, serve a similar role with many more under construction or planned. The high-speed services linking Zurich, Bern and Basel in Switzerland have brought the Central Business Districts (CBDs) of these three cities within 1 hour of each other. This has resulted in unexpectedly high demand for new commuter trips between the three cities and a corresponding increase in suburban rail passengers accessing the high-speed services at the main city-centre stations (or Hauptbahnhof).

A double-decker Cityrail train at Central Railway Station, Sydney

Train types

Commuter/suburban trains are usually optimized for maximum passenger volume, in most cases without sacrificing too much comfort and luggage space, though they seldom have all the amenities of long-distance trains. Cars may be single- or double-level, and aim to provide seating for all. Compared to intercity trains, they have less space, fewer amenities and limited baggage areas.

A KTM Komuter electric multiple unit at a station in suburban Malaysia.

Multiple unit type

Commuter rail trains are usually composed of multiple units, which are self-propelled, bidirectional, articulated passenger rail cars with driving motors on each (or every other) bogie. Depending on local circumstances and tradition they may be powered either by diesel engines located below the passenger compartment (diesel multiple units) or by electricity picked up from third rails or overhead lines (electric multiple units). Multiple units are almost invariably equipped with control cabs at both ends, which is why such units are so frequently used to provide commuter services, due to the associated short turn-around time.

A Linee S TSR double-decker train operate in Italy. Ferrovie Nord Milano.

Locomotive hauled services

Locomotive hauled services are used in some countries or locations. This is often a case of asset sweating, by using a single large combined fleet for intercity and regional services. Loco hauled services are usually run in push-pull formation, that is, the train can run with the locomotive at the "front" or "rear" of the train (pushing or pulling). Trains are often equipped with a control cab at the other end of the train from the locomotive, allowing the train operator to operate the train from either end. The motive power for locomotive-hauled commuter trains may be either electric or Diesel-electric, although some countries, such as Germany and some of the former Soviet-bloc countries, also use diesel-hydraulic locomotives.

A Metra push-pull locomotive commuter train approaches a platform in Deerfield, a suburb of Chicago.

Seat plans

In the U.S. and some other countries, a three-and-two seat plan is used. However, few people sit in the middle seat on these trains because they feel crowded and uncomfortable.[1] It is said one industrial designer for one of New York City's commuter railroads, Metro-North, told people: "I designed the aisle seat with a half-back and no upholstery, so it will be very uncomfortable to sit there. They'll move in and take the center seat!"[2] (This seating design can also be found on older New Jersey Transit and Long Island Rail Road rolling stock.)

In Japan, longitudinal (sideways window-lining) seat is widely used in some commuter rail to increase capacity in rush hours. Carriages are not organized to increase seating capacity even in the case of commuting longer than 50 km and commuters in Greater Tokyo Area have to stand in the train for more than an hour.

Commuter rail in the world

Africa

Currently there are not many examples of commuter rail in Africa. Metrorail operates in the major cities of South Africa, and there are some commuter rail services in Algeria, Morocco and Tunisia. In Algeria, SNTF operates commuter-rail lines between the capital Algiers and its southern and eastern suburbs. They also serve to connect Algiers' main universities to each other .

Asia

In Japan, commuter rail systems have extensive network and frequent service, and are heavily used. It is notable that many of them are run by private railway companies.

Yamanote Line which is representative of the Japanese commuter rail.

In India, commuter rail systems are present in major cities. Mumbai Suburban Railway, the oldest suburban rail system in Asia, carries more than 6.9 million commuters on a daily basis which constitutes more than half of the total daily passenger capacity of the Indian Railways itself. Kolkata Suburban Railway is huge and extensive and covers large areas in Kolkata's hinterland. The Chennai Suburban Railway is another railway of comparison. In Hyderabad, the MMTS mainly transports people from the city centre to HI-TEC city, the city's Information Technology hub. Other commuter railways in India include Delhi Suburban Railway, Pune Suburban Railway and Lucknow-Kanpur Suburban Railway.

Commuter trains are currently uncommon in China, although a small system has been inaugurated in Beijing in 2008. Commuter rail systems are also planned for Shanghai, Nanjing, Tianjin and around the Pearl River Delta region.

In Iran, SYSTRA has done a "Tehran long term urban rail study". SYSTRA proposed 4 express lines which are similar to RER suburban lines in Paris. Tehran Metro is going to construct express lines. For instance, the Rahyab Behineh, a consultant for Tehran Metro, is studying Tehran Express Line 2. Tehran Metro currently has a commuter line between Tehran and Karaj.

Other examples in Asia include Seoul Metropolitan Subway which has suburban lines operated by Korail in South Korea, KTM Komuter in Malaysia, The Philippine National Railways orange line in Metro Manila, Philippines and, KRL Jabotabek in Jakarta Metropolitan area, Indonesia.

Commuter train of Jakata Metropolitan Area.

Europe

Major metropolitan areas in most European countries are usually served by extensive commuter/suburban rail systems. Well-known examples include Beovoz in Belgrade, Serbia, S-Bahn in Germany, Switzerland and Austria, Proastiakos in Greece, RER in France, Linee S in Milan (Italy), Cercanías in Spain and HÉV in Budapest, Hungary.

In Russia, Ukraine and some other countries of the former Soviet Union, electrical multiple unit passenger suburban trains called Elektrichka are widespread.

Type X60 at Stockholm Central in Sweden

North America

In the United States, Canada, and Mexico regional passenger rail services are provided by governmental or quasi-governmental agencies, with a limited number of metropolitan areas served.

A Z 20500 train from Paris's RER line D running an old "ZYCK" (now renamed "ZUCO") route towards Melun

South America

Examples include an 899 km (559 mi) commuter system in the Buenos Aires metropolitan area, the 225 km (140 mi) long Supervia in Rio de Janeiro, and the Metrotrén in Santiago, Chile. Another example is Companhia Paulista de Trens Metropolitanos (CPTM) in Greater São Paulo, Brazil. CPTM has 93 stations with six lines, numbered starting on 7 (the lines 1 to 6 belong to the São Paulo Metro), with a total length of 260.8 kilometres (162.1 mi).

The Sarmiento Line is part of the extensive Buenos Aires metropolitan rail system.

Oceania

Major cities in Australia have suburban railway systems in their metropolitan areas. As with Japanese suburban railways or Germany's and Switzerland's S-Bahns, these Australian networks have far more frequent services and far higher ridership per capita than US 'commuter rail' in the usual sense of the term. To some US observers these networks may appear to operate as commuter/metro hybrids, however, they are simply the result of full utilisation of the available track capacity in the core of the suburban rail network, this being fed from multiple feeder lines. This is particularly so in Sydney and Melbourne, where headways on many lines in the core of the network reach 3–5 minutes in peaks and 10–20 minutes off peak (about 18 hours a day) and enter an underground loop for passenger distribution in the city centre; and where ridership per capita exceeds the sum of metro and commuter rail in comparable North American urban areas such as Toronto, Boston or the San Francisco Bay Area. All systems, however, are based on established main line rail systems, and track sharing with inter-city and freight services on parts of the network inhibit higher frequencies on some tracks. The main systems include:

- CityRail in Sydney, Newcastle and Wollongong. (largest metropolitan-wide network in Australia)
- Metro Trains Melbourne in Melbourne. ("Metro" being a technical misnoma)
- Queensland Rail in Brisbane.
- Transperth Trains in Perth.
- TransAdelaide in Adelaide.

New Zealand has two commuter rail systems: in Auckland and in Wellington (Tranz Metro).

See also

- List of suburban and commuter rail systems
- Public transport
- Commuting
- Cercanías, the commuter rail systems of Spain's major metropolitan areas
- Commuter rail in the United Kingdom
- Commuter rail in North America
- Cityrail, responsible for providing commuter rail services in and around Sydney, Newcastle and Wollongong, Australia
- S-Bahn, the combined city center and suburban railway system metro in Austria, Germany, Switzerland and Denmark

References

[1] (http://www.nytimes.com/2005/05/31/nyregion/31seat.html?pagewanted=1)
[2] (http://www.nytimes.com/2005/06/06/opinion/l06train.html)

External links

- Commuter Rail & Transit News (http://www.railserve.com/railnews/commuter_rail_transit.html) Current news concerning commuter rail development and issues

East Berlin

Ostberlin / Berlin (Ost) **East Berlin**

Soviet-occupied sector of Berlin,
East Germany

← 1949–1990 →

Flag

Coat of arms

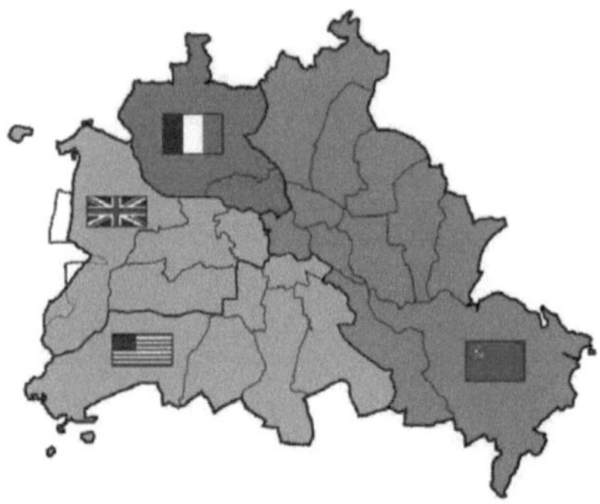

The four occupation zones of Berlin. East Berlin is in red

Historical era	Cold War
- Established	1949
- Reunification	3 October 1990
Area	
- 1989	409 km^2 (158 sq mi)
Population	
- 1989	1279212
Density	3127.7 /km^2 (8100.6 /sq mi)

East Berlin was the name given to the eastern part of Berlin between 1949 and 1990. It consisted of the Soviet sector of Berlin that was established in 1945. The American, British and French sectors became West Berlin, a part strongly associated with West Germany but a free city. Despite its status as part of an occupied city, East Berlin was claimed as the capital of East Germany, as the Soviets did not honor the distinction of a free city for their sector. From 13 August 1961 until 9 November 1989 East Berlin was separated from West Berlin by the Berlin Wall. The East German government referred to East Berlin simply as "Berlin" or often *"Berlin, Hauptstadt der DDR"* (Berlin, capital of the GDR). The term "Democratic Sector" was also used until the 1960s. (See also Naming conventions).

History

Overview

The Western Allies (the USA, Britain and France) never formally acknowledged the authority of the East German government to govern East Berlin; the official Allied protocol recognized only the authority of the Soviet Union in East Berlin in accordance with the occupation status of Berlin as a whole. The United States Command Berlin, for example, published detailed instructions for U.S. military and civilian personnel wishing to visit East Berlin.[1] In fact, the three Western commandants regularly protested the presence of the East German National People's Army (NVA) in East Berlin, particularly on the occasion of military parades. Nevertheless, the three Western Allies eventually established embassies in East Berlin in the 1970s, although they never recognized it as the capital of East Germany. Treaties instead used terms such as "seat of government."

On 3 October 1990, West and East Germany were reunited, thus formally ending the existence of East Berlin.

East Berlin today

Since reunification, the German government has spent vast amounts of money on reintegrating the two halves of the city and bringing services and infrastructure in the former East Berlin up to the standard established in West Berlin. Despite this, there are still obvious differences between eastern and western Berlin. Eastern Berlin has a distinctly different visual aspect, partly because of the greater survival of prewar façades and streetscapes, some still showing signs of wartime damage, and partly because of the distinctive style of urban Stalinist architecture used in the GDR. As in other former East German cities, a small number of GDR-era names commemorating socialist heroes have been preserved, such as Karl-Marx-Allee, Rosa-Luxemburg-Platz and Karl-Liebknecht-Straße; this followed a long process of review in which many such street names were deemed inappropriate and were changed. Still visible throughout former East Berlin are the characteristic "Ampelmännchen" on some pedestrian traffic lights. These days they are also visible in parts of the former West Berlin following a civic debate about whether the "Ampelmännchen" should be abolished or disseminated more widely.

Soviet and East German Commandants of East Berlin

Marx-Engels-Platz and the Palast der Republik in East Berlin in the summer of 1989. The Fernsehturm (TV Tower) is visible in the background

Boroughs of East Berlin (as of 1987)

Name [2]	Term
Nikolay Berzarin	2 May 1945 – 16 June 1945
Aleksandr Gorbatov	17 June 1945 – 19 November 1945
Dimitry Smirnov	19 November 1945 – 1 April 1946
Aleksandr Kotikov	1 April 1946 – 7 June 1950
Sergey Dienghin	7 June 1950 – April 1953
Pavel Dibrov	April 1953 – 23 June 1956
Andrey Chamov	28 June 1956 – 26 February 1958
Nikolay Zakharov	26 February 1958 – 9 May 1961
Andrey Soloviev	9 May 1961 – 22 August 1962
Helmut Poppe	22 August 1962 – 31 May 1971
Artur Kunath	1 June 1971 – 31 August 1978
Karl-Heinz Drews	1 September 1978 – 31 December 1988
Wolfgang Dombrowski	1 January 1989 – 30 September 1990
Detlef Wendorf	1 October 1990 – 2 October 1990

Boroughs of East Berlin

At the time of German reunification, East Berlin comprised the boroughs of

- Friedrichshain
- Hellersdorf (since 1986)
- Hohenschönhausen (since 1985)
- Köpenick
- Lichtenberg
- Marzahn (since 1979)
- Mitte
- Pankow
- Prenzlauer Berg
- Treptow
- Weissensee

Images of East Berlin

Karl Marx Allee apartments

Wall plaque of Lenin, off Wilhelmstraße

GDR-era mural of Meissen porcelain on former Council of Ministers building, facing Leipziger Straße

The Soviet War Memorial in Treptower Park

Cafe Moskau in Karl Marx Allee

The Palace of the Republic, being dismantled

New Synagogue, Oranienburger Straße

"Hochhaus" in Weberwiese- the first high rise apartment that was built after the war

People's Theatre, Rosa-Luxemburg-Platz

Late-1980s GDR apartment blocks on the Wilhelmstraße

Strausberger Platz with constructivism style building

Proletarian hero, Alexanderplatz

See also

- History of Berlin
- Berlin Wall
- Bonn
- Cold War
- Checkpoint Charlie
- East Germany
- Ghost station
- History of Germany since 1945
- West Berlin

Statues of Marx and Engels, Marx-Engels-Forum

References

[1] (PDF) *Helpful Hints for US Visitors to East Berlin* (http://upload.wikimedia.org/wikipedia/commons/7/7d/USCOB_Hints_on_Visiting_East_Berlin_-_19811109.pdf). Headquarters, U.S. Command Berlin. 1981-11-09. .

[2] "Commandants of Berlin Soviet Zone" (http://www.worldstatesmen.org/German_States1918.htm#Soviet). *World Statesmen.org*. .

External links

- Berlin Photos 1989–1999 (http://www.berlinphotos.co.uk)
- East Berlin Past and Present (http://www.ostberlin.de/en/)
- Pictures of the GDR and East Berlin 1949–1973 (http://www.ddr-fotos.de/gdr_photos.htm)
- Old East Berlin Fades Away Amid Renovations (http://www.npr.org/templates/story/story.php?storyId=6389848)
- The Lives of Others official website (http://www.thelivesofothers.com)

Friedrichshain

Friedrichshain

Quarter of Berlin

Oberbaumbrücke

Friedrichshain

Coordinates	52°30′57″N 13°27′15″E
Administration	
Country	Germany
State	Berlin
City	Berlin
Borough	Friedrichshain-Kreuzberg
Quarter subdivisions	3 zones
Basic statistics	
Area	9.78 km^2 (3.78 sq mi)
Elevation	52 m (171 ft)
Population	114050 *(30 June 2009)*
- Density	11662 /km^2 (30203 /sq mi)

Founded	1920
Other information	
Time zone	CET/CEST (UTC+1/+2)
Licence plate	B
Postal codes	(nr. 0201) 10243, 10245, 10247, 10249

Friedrichshain is a part of Berlin's borough of Friedrichshain-Kreuzberg, and like Kreuzberg across the river it has its own distinct character, with the result that the new double name is hardly ever used outside government administration. From its creation in 1920 until Berlin's 2001 administrative reform it was a freestanding city borough. Formerly part of East Berlin, it is an inner city locality, adjacent to Mitte, Prenzlauer Berg, Kreuzberg and Lichtenberg. Friedrichshain is named after the *Volkspark Friedrichshain*, a vast green park located at the north border to Prenzlauer Berg. During the Nazi era, the borough bore the name *Horst-Wessel-Stadt.*

Geography

Subdivision

Friedrichshain counts 3 zones (*Ortslagen*):

- Boxhagen
- Oberbaum-City
- Stralau

History

Karl-Marx-Allee, the broad boulevard that bisects Friedrichshain, seen from TV tower

The largely working-class district was created in 1920 when Greater Berlin was established by referendum, incorporating several surrounding settlements. Friedrichshain united the Frankfurter Vorstadt, already part of Berlin, and the villages of Boxhagen and Stralau. It took its name (meaning 'Frederick's Grove') from the *Volkspark* ('People's Park'), which was planned in 1840 to commemorate the centenary of Frederick the Great's coronation. Much of the district was settled in the rapid industrialization of the 19th and early 20th centuries, led by growth in manufacturing and crafts. It owed much to the opening of the railway line between Berlin and Frankfurt (Oder) in 1846 (which terminated near the site of today's Berlin Ostbahnhof), and the opening of the first waterworks in 1865 at Stralauer Tor. In 1874 the *Krankenhaus im Friedrichshain* was opened, Berlin's first hospital beside the university clinic Charité. In the early 1900s, the district's largest employer was the Knorr-Bremse brake factory; the Knorrpromenade, one of Friedrichshain's most attractive streets, was built to house the management.

When the Nazis came to power in 1933, the district was renamed Horst-Wessel-Stadt after the street fighter and writer of the Nazi hymn whose slow death, after being shot by communists, in Friedrichshain hospital in 1930 was turned into a propaganda event by Joseph Goebbels.

During World War II Friedrichshain was one of the most badly damaged parts of Berlin, as Allied strategic bombers specifically targeted its industries. As late as the nineties, some buildings still displayed bullet holes from the intense house to house fighting during the Battle of Berlin. After the war ended, the boundary between the US and Soviet occupation sectors ran between Friedrichshain and Kreuzberg. This became a sealed border between East and West Berlin when the Berlin Wall was built in 1961.

The Stalinallee (previously Große Frankfurter Straße) was built in Friedrichshain in the late 1940s and early 1950s as a prestige project; the architecture of its 'workers' palaces' is strongly reminiscent of the ostentatious Soviet-era Moscow boulevards and is sometimes mockinlgy described as *Zuckerbäckerstil* ('wedding cake style'). It was in these construction works that the 1953 uprising had its origins, when increased work quotas led to protests that would spread throughout East Germany, and were only put down by armed Soviet intervention.

In the period of De-Stalinization following the Soviet leader's death, the boulevard was renamed Karl-Marx-Allee at one end and Frankfurter Allee at the other. From this time onwards, Friedrichshain often featured on East Berlin's cultural map: in 1962 the *Kosmos*, East Germany's largest cinema was opened, followed in 1981 by the country's most ambitious swimming and sports complex, the *Sport- und Erholungszentrum*. Neither of these buildings serve their original function today.

Lifestyle

Simon-Dach-Straße is the "heart" of Friedrichshain and one of the liveliest places in Berlin

In the course of the changes following the fall of the Berlin Wall in November 1989, the free elections in March 1990 and leading up to German reunification the following October, Friedrichshain began to develop a reputation as a young, dynamic district, thanks in part to low rents and the many empty apartments that also attracted the attention of West Berlin squatters. On 14 November that year, Friedrichshain experienced violent clashes when hundreds of squatters were forcefully evicted from houses in *Mainzer Straße* by police acting on the orders of the Senate of the recently united city, an act which would trigger the fall of the governing coalition when the Green Party withdrew in protest. In the following years further squatters were evicted under the hardline conservative Senator for the Interior, but others were able to formally take possession of the houses they lived in, and they remain a distinct (counter-)cultural influence in the district to this day.

Alongside the neighboring districts of Mitte, Prenzlauer Berg, and Kreuzberg, Friedrichshain is now considered one of Berlin's most fashionable areas, and is home to numerous design and media companies including MTV Central Europe. It is known for its many bars, clubs, pubs, and cafes, concentrated in the vicinity of Simon-Dach-Straße and Boxhagener Platz. There were numerous squats in Friedrichshain, particularly in Rigaer Straße. In contrast to the more gentrified and expensive districts of Prenzlauer Berg and Mitte, Friedrichshain has a slightly run-down atmosphere, and its lower rents following German reunification attracted students and artists. Nowadays numerous restoration works are under way and Friedrichshain is developing at a fast pace becoming more and more gentrified itself.

At the opposite end of the district, the *Volkspark Friedrichshain* is a large park serving the densely populated areas of Friedrichshain, and Prenzlauer Berg on the other side. Its distinctive features include the *Märchenbrunnen* (Fairytale Fountain) and two green and pleasant "mountains" consisting purely of rubble and the ruins of two World War II Flak towers.

Points of interest

- East Side Gallery, a part of the Berlin Wall that was turned into an international outdoor gallery
- Karl-Marx-Allee, a boulevard lined with buildings in the Stalinist style, originally called *Stalinallee*
- Frankfurter Tor, two landmark towers on Karl-Marx-Allee, which resemble the church domes on Gendarmenmarkt
- Oberbaumbrücke, a road and rail bridge connecting Kreuzberg and Friedrichshain built in North German brick
- Fairytale Fountain in Volkspark Friedrichshain
- Simon-Dach-Straße, street with numerous pubs
- Boxhagener Platz, heart of the Friedrichshain *Kiez* or neighborhood
- Straße der Pariser Kommune, a street beginning a block north of the Karl-Marx-Allee and extending to the banks of the Spree
- Samariterviertel with the Samariterkirche (Church of the Good Samaritan)
- Berlin Ostbahnhof
- Volkspark Friedrichshain, with its Memorial to Polish Soldiers and German Anti-Fascists
- Skatehall Berlin, Revaler Strasse 99

Gallery

Simon-Dach-Straße is a popular destination, with numerous bars.

Jugendstil house in the Simon-Dach-Straße.

The Knorrpromenade.

The small cinema "Intimes," over 70 years old, on the corner of Boxhagener Straße.

A popular flea market takes place every Sunday on Boxhagener Platz.

Frankfurter Allee, looking west, with the Fernsehturm at Alexanderplatz in the background.

The north tower at Frankfurter Tor.

A closer look at a Frankfurt Allee façade, showing the neo-classicist tilework.

The Haus an der Weberwiese, the first high-rise apartment block in Berlin, built 1951.

The "Molecule man", a big art work at the Berlin river Spree.

The Osthafen (east harbour) at the Spree between Friedrichshain and Kreuzberg. In the background the Oberbaumbrücke and the Fernsehturm.

The Oberbaumbrücke and the waters of the Spree belong to Friedrichshain.

The East Side Gallery is the longest remaining part of the Berlin Wall.

A squat on Kreuziger Straße, one of many such buildings in the district.

See also

- Flakturm II - Friedrichshain
- Berlin Friedrichshain-Kreuzberg - Prenzlauer Berg East (electoral constituency)
- List of streets and squares in Berlin-Friedrichshain

External links

- - Official Berlin Parks Website [1]
- Die Friedrichshainer - unofficial city portal [2]
- siebenschoen - a true toy story [3]
- Amitola - The Children Shop [4]
- fensterzumhof.eu: Photos of Friedrichshain [5]
- Apartments Am Boxi: Affordable Apartments with views over the Boxhagener Platz [6]
- Friedrichshain travel guide from Wikitravel

References

[1] http://www.stadtentwicklung.berlin.de/umwelt/berlin_move/en/laufstrecken/laufstrecke11.shtml

[2] http://www.die-friedrichshainer.de

[3] http://siebenschoen-berlin.com

[4] http://www.amitola-berlin.de

[5] http://www.fensterzumhof.eu/bilder/key/Friedrichshain

[6] http://en.fewo-berlin.biz

Friedrichshain-Kreuzberg

Arms	Map

Area:	20.16 km²
Inhabitants:	268,225 *(2010-03-31)*
Population density:	13,304 inhabitants per km²
Website:	official homepage [1] (German)
Politics	
Mayor of Borough:	Franz Schulz (Alliance '90/The Greens)
Parliament of Borough (*BVV*):	Alliance '90/The Greens 22, SPD 13, The Left 7, Pirates 5, CDU 4 (last election 2011)

Friedrichshain-Kreuzberg is the second borough of Berlin, formed in 2001 by merging the former East Berlin borough of Friedrichshain and the former West Berlin borough of Kreuzberg. The historic Oberbaum Bridge, formerly a Berlin border crossing for pedestrians, links both districts across the river Spree as the new borough's landmark (as featured in the coat of arms).

Overview

The counterculture tradition especially of Kreuzberg led to a plurality of votes for the Green Party, which is unique among all Berlin boroughs. The local MP Hans-Christian Ströbele is the only Green politician directly elected to the federal Bundestag. While Kreuberg is characterised by a high number of immigrants, the share of non-German citizens in Friedrichshain is much lower and the average age is higher. The merger between the distinct quarters is celebrated by an annual anarchic "vegetable fight" on the Oberbaumbrücke. Both parts have to deal with the consequences of gentrification.

Subdivision

Friedrichshain-Kreuzberg is divided into 2 localities:

- Friedrichshain
- Kreuzberg

Twin towns

- Wiesbaden, Germany since 1964
- Porta Westfalica since 1968
- Bergstraße district, since 1969
- Ingelheim am Rhein since 1971
- Limburg-Weilburg district, since 1980
- San Rafael del Sur, Nicaragua since 1986
- Kiryat Yam, Israel since 1990
- Szczecin, Poland since 1996
- Kadıköy, Istanbul, Turkey since 1996
- Oborishte, Sofia since 1999

See also

- Berlin Friedrichshain-Kreuzberg - Prenzlauer Berg East (electoral constituency)

External links

- Official homepage (German)[[Category:Articles with German language external links [2]]]
- Official homepage of Berlin [3]
- Xhain.info - Private portal (German)[[Category:Articles with German language external links [4]]]

References

[1] http://www.friedrichshain-kreuzberg.de
[2] http://www.friedrichshain-kreuzberg.de/
[3] http://www.berlin.de/english/
[4] http://www.xhain.info/

Rummelsburg

Rummelsburg	
Quarter of Berlin	

Tuchollaplatz (Victoriastadt)

Rummelsburg

Coordinates	52°30′0″N 13°29′33″E
Administration	
Country	Germany
State	Berlin
City	Berlin
Borough	Lichtenberg
Quarter subdivisions	1 zone
Basic statistics	
Area	4.52 km^2 (1.75 sq mi)
Elevation	50 m (164 ft)
Population	17567 *(30 June 2008)*
- Density	3887 /km^2 (10066 /sq mi)

Founded	1669
Other information	
Time zone	CET/CEST (UTC+1/+2)
Licence plate	B
Postal codes	(nr. 1112) 10317

Rummelsburg is a German locality (*Ortsteil*) in the borough (*Bezirk*) of Lichtenberg, Berlin.

History

The locality was founded in 1669 and on January 30, 1889 became a rural municipality, with the name of Boxhagen-Rummelsburg. Merged in 1912 in the town of Lichtenberg, in 1920 it was incorporated in Berlin with the "Greater Berlin Act" [1].

Geography

Overview

Located in the eastern side of the city and crossed by Spree river in the western corner, Rummelsburg borders with Lichtenberg, Friedrichsfelde, Karlshorst, Friedrichshain (in Friedrichshain-Kreuzberg district), Plänterwald and Oberschöneweide (both in Treptow-Köpenick district). The lake Rummelsburger See belongs to the locality and separates it from Stralau, a zone of Friedrichshain.

Subdivision

Rummelsburg counts 1 zone (*Ortslage*):

- Victoriastadt

Transport

The locality is served by several urban rail lines of *S-Bahn* and *U-Bahn*. The stations serving the locality are Rummelsburg (S3 line), Betriebsbahnhof Rummelsburg (S3), Nöldnerplatz (S5, S7, S75), partly Ostkreuz (S3, S4, S5, S7, S75, S8, S85, S9) and part of the DB station of Lichtenberg (S5, S7, S75, U5, Tram).

Personalities

- Adolph Schlicht (1840–1910)
- Heinrich Zille (1858–1929)

Photogallery

Nöldnerplatz

The Rummelsburger See

Münsterlandplatz

Klingenberg power station

References

[1] (German) Historical infos about Rummelsburg (http://www.luise-berlin.de/strassen/bez17h/r640.htm)

External links

- (German) Rummelsburg page on www.berlin.de (http://www.berlin.de/ba-lichtenberg/derbezirk/rummelsburg.html)

Berlin Westkreuz station

DB	
Berlin-Westkreuz	

Sole entrance of Westkreuz station	
Operations	
Category	4
Type	Bf
DS100 code	BWKR
Station code	6723
Construction and location	
Location	Berlin
State	Berlin
Country	Germany
Local authority	Charlottenburg-Wilmersdorf
Am Westkreuz 14055 Berlin	
Route information	
• Berlin Stadtbahn • Ringbahn	
List of railway stations in the Berlin area	

Berlin Westkreuz is a station in the Charlottenburg district of Berlin. It is served by the S-Bahn lines S3, S41, S42, S46, S5, S7 and S75 and so represents a major interchange point on the Berlin S-Bahn network. It lies at the opposite end of the Stadtbahn to Ostkreuz and is one of the four main stations on the Ringbahn (circle line).

History

Westkreuz station was first opened on 10 December 1928, as a two-level-station under the name *Ausstellung* (Exhibition), which was changed to its current name in 1932. In 1980, with the cessation of services on this portion of the Ringbahn, Westkreuz lost its function as an interchange and had only a minimal service on the Stadtbahn. The area surrounding the station is very sparsely populated and this was its major purpose, making it a very poorly-frequented station during this period.

However, after the fall of the Berlin Wall, services were gradually reinstated and on 17 December 1993, the Ringbahn was reopened between Westend and Baumschulenweg, restoring the station's significance as an interchange point. In the following years, the station was entirely refurbished and is now one of the most modern on the network.

S7 train en route for Ahrensfelde arriving at the lower level Westkreuz station

Preceding station	Berlin S-Bahn	Following station
Messe Süd *toward Spandau*	S3	Charlottenburg *toward Erkner*
Halensee *One-way operation*	S41	Messe Nord/ICC *toward Ringbahn (clockwise)*
Messe Nord/ICC *One-way operation*	S42	Halensee *toward Ringbahn (counter-clockwise)*
Messe Nord/ICC *toward Westend*	S46	Halensee *toward Königs Wusterhausen*
Terminus	S5	Charlottenburg *toward Strausberg Nord*
Grunewald *toward Potsdam Hbf*	S5 (night)	
	S7	Charlottenburg *toward Ahrensfelde*
Messe Süd *toward Spandau*	S75	Charlottenburg *toward Wartenberg*

Berlin Stadtbahn

Stadtbahn	
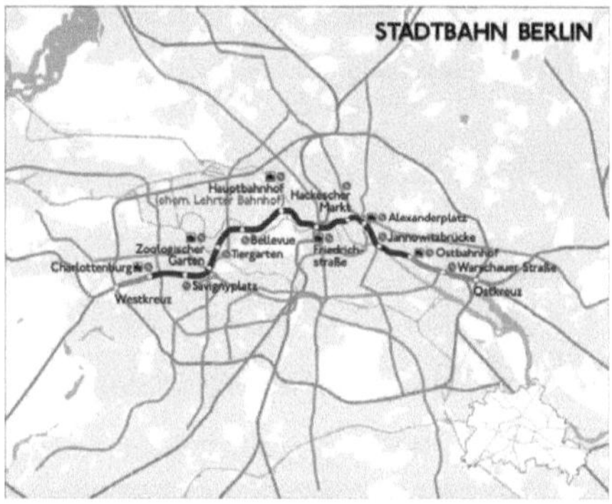	
Line number:	6003, 6024
Track gauge:	1435 mm (4 ft 8 $^1/_2$ in)
	Legend

km	Station
	from Lichtenberg / Karlshorst
0.0	Ostbahnhof
1.1	Jannowitzbrücke U8
2.2	Alexanderplatz U2 U5 U8
2.9	Hackescher Markt
4.0	Friedrichstraße Nord-Süd-Tunnel S1 S2 S25 U6
5.4	Hauptbahnhof North–South mainline U55
7.1	Bellevue
7.9	Tiergarten crossover
8.2	Tiergarten
9.0	Zoologischer Garten U2 U9
10.0	Savignyplatz
11.2	Charlottenburg U7
	North/south curve to Ring line
	to Westkreuz

The **Berlin Stadtbahn** ("city railway") is a major railway thoroughfare in the German capital Berlin, which runs through Berlin from east to west. It connects the eastern district of Friedrichshain with Charlottenburg in the west via 11 intermediate stations including Hauptbahnhof. The Berlin Stadtbahn is often also defined as the slightly longer route between Ostkreuz and Westkreuz, although this is not technically correct.

The line was originally built in the 1880s. It is 12 kilometers (7.5 mi) in length, and is entirely elevated above the city's streets. The four track route carries S-Bahn, Regionalbahn, Regional-Express, Intercity, EuroCity and

Intercity-Express trains.

Operation

Structure and tracks

The Stadtbahn line is an elevated rail line with viaducts totalling 8 kilometers (5.0 mi) in length and including 731 masonry viaduct arches. A further 2 kilometers (1.2 mi) of the line are situated on 64 bridges, that cross adjoining streets and (three times) the River Spree. The remaining length of the line is on an embankment.

The line carries four tracks, in two pairs. The northern pair are reserved for use by the S-Bahn, and are electrifed using a third rail carrying 800V DC. The S-Bahn tracks have platforms at all eleven stations along the Stadtbahn.[1]

The southern pair of tracks are used by Regionalbahn, Regional-Express, Intercity, EuroCity and Intercity-Express trains, and are electrified using the German standard of 15kV at 16,7Hz AC, supplied by overhead line. Six of the Stadtbahn stations have platforms on these tracks, although not all trains stop at all stations, depending on the class and route of the train.[1]

Stations

From east to west, the Stadtbahn has stations at:[1]

- Ostbahnhof
- Jannowitzbrücke (S-Bahn only)
- Alexanderplatz
- Hackescher Markt (S-Bahn only)
- Friedrichstraße
- Hauptbahnhof
- Bellevue (S-Bahn only)
- Tiergarten (S-Bahn only)
- Zoologischer Garten
- Savignyplatz (S-Bahn only)
- Charlottenburg

Hackescher Markt station, with RegionalExpress and S-Bahn trains

Routes

The S-Bahn tracks of the Stadtbahn currently carry routes S3 (Erkner to Spandau), S5 (Strausberg Nord to Westkreuz), S7 (Ahrensfelde to Potsdam) and S75 (Wartenberg to Spandau).[2]

The longer distance tracks carry Regionalbahn and Regional-Express routes RE1 (Magdeburg to Eisenhüttenstadt), RE2 (Rathenow to Cottbus), RE7 (Dessau to Wünsdorf-Waldstadt) and RB14 (Nauen to Berlin Schönefeld Airport). Although most InterCity and Intercity-Express trains now use the north-south tunnel route via Hauptbahnhof, some trains do still remain on the Stadtbahn's long distance tracks. These trains, mainly those heading toward Hanover and Cologne, usually call at Hauptbahnhof and Ostbahnhof.[3] [4]

History

Planning

In 1871, eight main line railways existed in Berlin, each with its own terminal station at the city's edge or outside the city limits. This was very impractical for many passengers, who were forced to use hackney carriages to transfer from one train to another. Therefore a railway line was planned to connect these terminuses with each other.

In 1872, the *Deutsche Eisenbahnbaugesellschaft* (German Railway Construction Company - DEG) filed the planning application for a railway line through the city, connecting the then-Schlesischer Bahnhof (today Berlin Ostbahnhof) to Charlottenburg, and continuing to Potsdam. In December 1873, the state of Prussia as well as the private rail enterprises Berlin-Potsdamer Eisenbahn, Magdeburg-Halberstädter Eisenbahn and Berlin-Hamburger Bahn bought shares in the DEG, and jointly founded the *Berliner Stadteisenbahngesellschaft* (Berlin City Railway Company). However, things did not go as expected and the DEG went into bankruptcy in 1878, which forced the Prussian state government to take over operations, pay for the construction of the line with state money and to reimburse the former private owners of the DEG.[5] The state's interest in the line was attributed to the military, which after the 1870-1871 Franco-Prussian War was of the opinion that the railway networks would hinder mobilisation when not properly interconnected.[6]

On July 15, 1878 the *Königliche Direktion der Berliner Stadteisenbahn* (Royal Directorate of Berlin City Railways), under the management of Ernst Dircksen, was commissioned to manage the site. The directorate at first reported to the Prussian Ministry of Transport and later became a subsidiary of the Ministry of Public Operations.

Stadtbahn seen from the Fernsehturm, near station Hackescher Markt

The planned railway had two tracks each for freight and passenger traffic. Having taken similar projects in London and New York City into consideration, passenger traffic received priority over freight trains. Furthermore, the new railway line was not only to serve as a connection between the mainline terminuses in Berlin, but would also offer connections to the Berlin Ringbahn and the suburban rail lines.

The traffic routing was not only influenced by the location of the already existing stations the line was supposed to connect, but also by land availability in the city centre. One of the original drafts, which called for building the line along Leipziger Straße, had to be scrapped because of overly high land prices. The moat of the 17th century Berlin Fortress was filled up between today's stations Hackescher Markt and Jannowitzbrücke and, since it was public land, was used for building the railway line. This explains some of the curvy sections on the Stadtbahn, especially between the stations Alexanderplatz and Jannowitzbrücke.[7]

Berliner Stadteisenbahn: Station Savignyplatz with a Prussian T2

Its elevated nature sets the Stadtbahn apart from the previous Berliner Verbindungsbahn, built in 1851, which was built at street level and was a hindrance to travel. The Stadtbahn was originally equipped with vertical iron sleepers, however these were replaced with wooden sleepers in the early 20th century.

Construction

Work on the line started in 1875, and the Stadtbahn was opened on February 7, 1882 for local traffic; it opened on May 15 of the same year for long-distance trains. The costs of construction, including purchase of the land, were estimated at about 5 million Goldmark per kilometre. The line would later become the core route of the Berlin S-Bahn.

The original stations, seen from west to east, were:

- Charlottenburg
- Zoologischer Garten
- Bellevue
- Lehrter Stadtbahnhof, today Berlin Hauptbahnhof
- Friedrichstraße
- Börse, Marx-Engels-Platz in the GDR, since 1992 Hackescher Markt
- Alexanderplatz
- Jannowitzbrücke
- Schlesischer Bahnhof, named Berlin Hauptbahnhof in the 1980s and early 90s, today Ostbahnhof

Since May 1, 1888 the Stadtbahn also connected to Stralau-Rummelsburg (Ostkreuz since 1933) in the east and Westend (via Westkreuz) in the west.

Two stations were added between Charlottenburg and Zoologischer Garten as well as between Zoologischer Garten and Bellevue:

- Savignyplatz (August 1, 1896)
- Tiergarten (January 5, 1885)

Initial operation

Suburban trains

Suburban trains operated on the local tracks, the so-called *city track.* At first, these were either services to the suburbs or connections to the Berlin Ringbahn, running as "half ring trains", using the Stadtbahn and either the northern or the southern Ringbahn. The trains were pulled by locomotives, which ran on coke to minimise the smell. Doors on the train compartments had to be opened by the passengers themselves and stations were not called out on the train.

These trains ran from 4 o'clock in the morning to 1 o'clock at night, typically at intervals between two and five minutes, depending on the time of day. Fares in the early 20th century were 10 pfennig in 3rd class and 15 pfennig in 2nd class.

Freight traffic

The freight traffic to the central market at Alexanderplatz was carried by seven special trains per day, of which four ran at night, two during the day and one in the evening. Furthermore, the line carried the normal freight traffic toward Charlottenburg, Moabit, Wedding, Zentralviehhof, Weißensee, Frankfurter Allee, Rixdorf, Tempelhof, Wilmersdorf-Friedenau and to Halensee.

Long-distance trains

In the first years of the Stadtbahn, many trains previously terminating at the old terminuses Lehrter Bahnhof, Görlitzer Bahnhof or Potsdamer Bahnhof operated via the Stadtbahn to reduce the load on the terminus stations. By the end of the 19th century, however, most of these train runs had to terminate at their old destination stations again due to the increasing local traffic on the Stadtbahn.

The remaining traffic on the Stadtbahn mostly consisted of express trains to Hanover and Cologne via the Lehrter Bahn, Kanonenbahn trains to Dessau, trains to Königsberg and Danzig on the Preußische Ostbahn and trains to Frankfurt/Oder and Breslau. Suburban trains to Spandau and Strausberg also ran on the Stadtbahn's long-distance tracks until 1928.

Trains heading west usually left from the Schlesischer Bahnhof station, those heading east from Charlottenburg. Depots were situated in Rummelsburg (then called *Bw Karlshorst*) and Grunewald.

Station expansions and viaducts

Station Jannowitzbrücke

In 1914, the Friedrichstraße station was rebuilt; the long-distance section of the station was expanded to four tracks and the current station hall was built. Between 1922 and 1932, the Stadtbahn viaduct was thoroughly modernised in order to handle the ever increasing train weight. Also, the train sheds of Alexanderplatz and Schlesischer Bahnhof were replaced. The suburban line's platforms were raised to a height of 96 centimetres.[8]

A second long-distance platform and a new hall were built at Zoo station from 1934 to 1940. The station hall was only glazed in the 1950s, however. The notable terraced vestibule dates from the same time.

Electric operation

The S-Bahn logo.

On June 11, 1928 the suburban line Potsdam-Stadtbahn-Erkner was fully equipped with DC third rail gear.[9] Five trains of the new DRG Class ET 165 – the type appropriately named *Stadtbahn* – went into service, still sharing the track with steam trains. By November 1928, all lines leading toward the Stadtbahn, namely the lines from Kaulsdorf, Spandau and Grünau as well as the Berlin Ringbahn, were fully electrified. Therefore, suburban services to Spandau could be moved from the long-distance tracks to the local tracks. The last steam trains disappeared in 1929 when the ring became fully operated by electric trains. Half-ring trains operated only as peak time services. In December 1930, the term ***S-Bahn*** and the symbol of a white S on a green circle were introduced for the city, ring and suburban services.

Post-War situation

After World War II, the Stadtbahn lay devastated by bombs, but was rebuilt very quickly. Because Stalin wanted to travel by train to the Potsdam Conference, the Stadtbahn was converted to the Russian 1524 mm broad gauge in 1945. The national importance had diminished with the loss of eastern Germany; only a few trains ran towards the western zones. Some trains from the Soviet zone terminated on the Stadtbahn.

During the Berlin Blockade, the long-distance traffic came to an almost complete halt. The Stadtbahn was useful for the re-established S-Bahn, however, now with connections to places line Königs Wusterhausen, Strausberg, Staaken and Falkensee.

From May 18, 1952, when all Berlin terminal stations and all other long-distance stations in West Berlin were closed, the station Zoologischer Garten remained the only long-distance station for the western part of the city. The last domestic train of the GDR (East German) railways ran on the Stadtbahn in 1953.

Transit train at Zoo station 1976

After the Berlin Wall was built in 1961, Zoologischer Garten became the West Berlin and Ostbahnhof the East Berlin *de facto* central station. The station Friedrichstraße now was the terminal point of the separated West and East Berlin S-Bahn lines and departure point for the Interzonenzug (Inter-zone train) services between West Berlin and West Germany. Friedrichstraße station was separated into Eastern and Western parts with steel walls, and enabled West Berliners to change to S-Bahn trains running on the Nord-Süd-Bahn and the U6 line of the Berlin U-Bahn without passing through GDR border controls. The station also featured a border crossing into East Berlin.[10] Through trains between Zoo and Ostbahnhof only existed in international traffic, for example the Paris to Warsaw trains. Later, through coaches and shuttle trains connecting to the night trains to Scandinavia crossed the intra-German border on the Stadtbahn as well.

Due to a quirk in legislation, the West Berlin parts of the Stadtbahn belonged to the Deutsche Reichsbahn, which made it (and therefore the GDR government) one of the largest landowners in West Berlin. Regular quarrels erupted between the DR, the GDR government, the West Berlin Senate and the Allied occupation powers.

Prices for the West Berlin S-Bahn were kept slightly below the fares of the West Berlin BVG. In East Berlin, a flat fee of 0.20 Mark was charged until 1991. West Berlin politics and most of the populace fully boycotted the S-Bahn, which was run by the East German railways, and introduced bus and U-Bahn lines running parallel to the S-Bahn network and the Stadtbahn.

Restoration during the Cold War

Despite problems, the Deutsche Reichsbahn made improvements to the line and reconstructed at great expense the Westkreuz railway station which had been built on swamp land.

Meanwhile, the number of S-Bahn lines running in West Berlin was reduced to just three as a consequence of a strike carried out by the Deutsche Reichsbahn's West Berlin-based employees in September 1980.

On January 9, 1984, a treaty between the GDR and the West Berlin Senate came into force and turned over the responsibility for operation of the S-Bahn in West Berlin to the West Berlin transport authority, the Berliner Verkehrsbetriebe. Soon thereafter, talks with the GDR commenced regarding improvements of the Stadtbahn in West Berlin as well as the modernisation of the Zoologischer Garten station. The Lehrter Stadtbahnhof was also carefully restored to its original 1880s look and became a listed building. In East Berlin, the Ostbahnhof was partially rebuilt and renamed to *Hauptbahnhof*, in time for the 750th anniversary of Berlin's founding in 1987.[11]

After reunification

The fall of the Berlin Wall and German reunification made for a sudden spike of importance for the Stadtbahn. The first InterRegio train ran to Cologne in 1990. Since 1991, Intercity trains to Karlsruhe, Cologne and Hamburg used the Stadtbahn. The Hamburg line was soon extended to Dresden and Prague, and the former Interzonenzug trains from Munich were converted to InterCity trains and now ran on the Stadtbahn as well.

The western part of the Stadtbahn was electrified on July 4, 1993 up to Zoologischer Garten station. The eastern part of the line up to Ostbahnhof had been electrified since 1987. As soon as electrification reached the Zoo station, ICE trains began to use the station.

Modernisation from 1994

In October 1994, a large-scale modernisation programme was started on the Stadtbahn. The viaducts were checked and strengthened, and the tracks were bedded in concrete to improve durability and comfort. Almost all stations saw large financial investments and were thoroughly modernised. Long-distance traffic between Zoo and Ostbahnhof stations was interrupted during the construction period and the S-Bahn trains temporarily used the long-distance line. On March 24, 1998 the Stadtbahn was reopened, now carrying up to three ICE and IC lines as well as five RegionalExpress lines.[12] Part of the Stadtbahn was realigned as part of the construction of the new Berlin Hauptbahnhof, with construction commencing in 2001 and completed in July 2002. The realigned section consisted of two 450 m-long bridges spanning the station and the adjacent Humboldthafen port.

Until the summer of 2006, the Stadtbahn was the main thoroughfare for long-distance trains, which usually stopped at Zoologischer Garten and Ostbahnhof (which was renamed in 1998). When the new Hauptbahnhof opened on May 28, 2006, the importance of the line diminished slightly, as many trains now would use the new north-south line connected to the Hauptbahnhof. The remaining intercity trains on the Stadtbahn, mainly those heading toward Hanover and Cologne, now usually call at Hauptbahnhof and Ostbahnhof.[4]

References

[1] (in German) *Eisenbahnatlas Deutschland [Railway Atlas for Germany]*. Verlag Schweers + Wall GmbH. 2009. pp. 128–29. ISBN 978-3-89494-139-0.

[2] "Stadtplan [City Map]" (http://www.s-bahn-berlin.de/stadtplan/) (in German). S-Bahn Berlin GmbH. . Retrieved May 4, 2011.

[3] "Liniennetz Regionalverkehr [Route map regional railways]" (http://www.vbbonline.de/download/pdf/liniennetze/100111_Bahnregionalverkehr.pdf) (in German) (PDF). Verkehrsverbund Berlin-Brandenburg. . Retrieved May 4, 2011.

[4] (in German) *Die Bahn am Ball*. Deutsche Bahn AG. 2006. pp. 130–36.

[5] "Die Berliner Stadtbahn" (in German). *Illustrirte Zeitung*. February 25, 1882.

[6] Elkins, Thomas Henry (1988). *Berlin: Spatial Structure of a Divided City*. Routledge. p. 114. ISBN 0416922201.

[7] "Die Linienführung der Berliner Stadteisenbahn" (in German). *Zeitschrift für Bauwesen*. 1884.

[8] "Fünfzig Jahre Berliner Stadtbahn" (in German). *Die Reichsbahn*. 1932.

[9] Bousset, Johannes (1935) (in German). *Die Berliner U-Bahn*. W. Ernst & Sohn. pp. 140.

[10] Katona, Marianna S. (2005). *Tales from the Berlin Wall. Recollections of Frequent Crossings.*. BoD. pp. 31–35. ISBN 3833404396.

[11] Deutsche Bahn AG (September 23, 2006). "Stations in Berlin, the railway capital" (http://web.archive.org/web/20070211125155/http://www.db.de/site/bahn/en/db__group/corporate__group/history/topics/berlins__stations/berlins__stations.html). Archived from the original (http://www.db.de/site/bahn/en/db__group/corporate__group/history/topics/berlins__stations/berlins__stations.html) on February 11, 2007. . Retrieved February 28, 2007.

[12] Wehling, Hans-Georg (2004) (in German). *Die deutschen Länder: Geschichte, Politik, Wirtschaft*. VS Verlag. p. 78. ISBN 353143229X.

Further reading

- (German) Die Stadtbahn, Signal-Sonderausgabe, Berliner S-Bahn-Museum, Berlin 1996, Verlag GVE, ISBN 3-89218-046-6

External links

- (German) Entry in the listed building directory of the Berlin Senate (http://www.stadtentwicklung.berlin.de/cgi-bin/hidaweb/getdoc.pl?LIST_TPL=lda_list.tpl;DOK_TPL=lda_doc.tpl;&KEY=obj 09011323)

This article incorporates information from the revision as of 2007-01-28 of the equivalent article on the German Wikipedia.

Berlin Ringbahn

The **Ringbahn** (German for: *Circular railway*) is a 37.5 km (23.3 mi) long railway line of the Berlin S-Bahn around the city centre of Berlin in Germany. The line is made up of the S-Bahn ring and the freight ring. S-Bahn service on the line is carried out by the S 41 (clockwise) and S 42 (counter-clockwise) circle lines, with 400,000 passengers a day.[1] Due to its distinctive shape, the line is colloquially called "Hundekopf" (dog's head).[2]

The Ringbahn also constitutes the outer border of the "A" zone for the Verkehrsverbund Berlin-Brandenburg transport association's fare structure, and is the outer border of the road traffic control zone for particulate matter control established by January 1 2008.

Ringbahn platform at Westkreuz

Ringbahn, Messe Nord ICC station

History

Background

In 1851 the Berliner Verbindungsbahn, also known as the *Königliche Bahnhofs-Verbindungsbahn* or *Royal Station Connection Raiway*, was completed between the terminal stations of the railways terminating in Berlin. This railway was laid directly in the streets of the urban area, disrupting traffic and also disturbing residents. In order to minimise the disruption of traffic, trains ran at night, with the train bell being constantly rung.

Therefore, plans were soon developed to build a new connecting line primarily for freight, which would run outside the then city limits. Funding for the construction was possible only after the victory in the war with Austria of 1866. Construction began in 1867 and the ring line was completed in 1877. The Lower Silesia-Mark Railway Company was commissioned to construct and manage the line.

Route

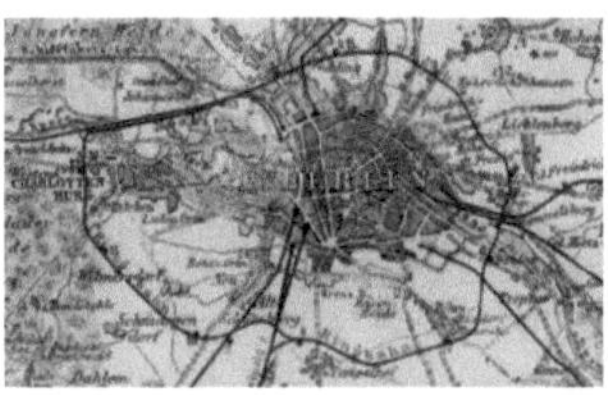

Ring Railway in 1885

The first section of the railway was opened on 17 July 1871 from the station of Moabit, through Gesundbrunnen, Central-Viehhof (now Storkower Straße), Stralau-Rummelsburg (now Ostkreuz), Rixdorf (now Neukölln) and Schöneberg (later Kolonnenstraße, now Julius-Leber-Brücke) to Potsdamer ring station, an annex to Potsdamer station. From there, trains returned in the opposite direction. This section into Potsdamer ring station became known as the *Südringspitzkehre* (*Southern ring zigzag*; this name reflected the need for trains to reverse there, eventually to continue their trip around the ring). The line crossed the Anhalt Railway (and later the Royal Prussian Military Railway) on bridges.

With the opening of the section from Schöneberg through the newly independent city of Charlottenburg (now Westend station) to Moabit on 15 November 1877 the ring was complete, with Potsdamer station still connected to the ring via the *Südringspitzkehre* for passenger trains.

In World War II the Potsdamer and Anhalter stations were heavily bombed, so since 1944 the *Südringspitzkehre* has been closed.

From 1944 until the construction of the Berlin Wall in 1961, S-Bahn trains ran over the direct line between the stations of Papestraße (now Südkreuz) and Schöneberg opened in 1933, making a complete circle. With the building of the Wall the line was broken in two places:

- In West Berlin a separate line on a *three-quarter ring* ran between Gesundbrunnen and Sonnenallee or Köllnische Heide.
- In East Berlin the remaining section of the ring line ran between Schönhauser Allee and Treptower Park, running on the suburban lines to Bernau and Königs Wusterhausen or Schönefeld Airport.

After the 1980 S-Bahn strike, services on the western part of the ring was suspended for about 13 years.

On January 9, 1984, a treaty between East Germany and the West Berlin Senate entered into force and turned over the responsibility for operation of the S-Bahn in West Berlin to the West Berlin transport authority BVG. It was initially planned to restore the section between Westend and Sonnenallee.

After German reunification in 1990 plans were changed so that in 1993 the south ring was reopened to the junction with the line towards Baumschulenweg along with that line to allow a connection with the Goerlitz line. The reconstruction of the connection between Sonnenallee and Treptow Park required large-scale renovation that was not feasible in the short term. In the following years, the western part of the ring line has been put back into operation in stages. In 2002, the S-Bahn ring line was fully restored. Since May 2006, a full circular service has been operated as lines S41 (clockwise) and S42 (anticlockwise).

Services

Starting from 1 January 1872 freight was carried out on the line to freight yards separate from the passenger stations. The operation of the ring line was electrified in 1926. In 1930, ring line operation were combined with the Stadtbahn and suburban services as the Berlin S-Bahn.

The building of the Berlin Wall in 1961 prevented continuous operation, after which passenger numbers on the West Berlin side, between Gesundbrunnen and Sonnenallee, continuously declined. This was caused partly by a politically-motivated call for a boycott, because revenue from the West Berlin S-Bahn, which were operated by East German railways, supported the East German government. In contrast, the East Berlin section of the ring from Schönhauser Allee to Treptow Park remained in operation as it formed part of a major north-south tangent.

The S-Bahn ring line operations on the western part of the line closed in 1980 resumed on 17 December 1993 on the section between Baumschulenweg, Neukölln and Westend. Further sections were subsequently reopened: Westend–Jungfernheide on 15 April 1997; Neukölln–Treptow Park on 19 December 1997 and Jungfernheide–Westhafen on 19 December 1999. On 17 September 2001 the S-Bahn returned to the route over the old border between Schönhauser Allee and Gesundbrunnen.

It was not until more than twelve years after the fall of the Wall, that the last gap of the S-Bahn between Westhafen, Wedding and Gesundbrunnen was restored on 16 June 2002. Promotional material for the reopening referred to this as "Wedding Day", in allusion to the English word wedding. Services at that time did not immediately return to a continuous circle operation. For the time being, operated under the "screw concept": trains entered the ring from the south at Neukölln and circled around it one and a half times until they ended at a station on the ring line. At the time the trip around the ring could not be achieved in less than 63 minutes.

Since 28 May 2006, the S-Bahn train have run on the ring line continuously. The trains take around 60 minutes with trains operating every five minutes in peak hour, and every ten minutes between the peaks and in the evenings. This is achieved through consistent use of the greatly accelerated 481/482 series trains. Some sections of the ring are used by other lines. On the southern ring coming from the Görlitz line in the southeast, S45 trains terminate at

Hermannstraße, S46 trains terminate at Westend and S47 trains terminate at Südkreuz, with some terminating at Bundesplatz. On the eastern section of the ring line S8, S85 and S9 trains operate between Schönhauser Allee and Treptower Park.

Under the so-called "mushroom concept" the long-distance lines on the northern part of the ring line for regional or long-distance services were rebuilt and electrified. On the ring line regional and mainline services now stop at Gesundbrunnen station and regional services stop at Jungfernheide.

The majority of the former ring line freight yards have been closed down or dismantled. Part of the former freight inner ring between Neukölln and Tempelhof is still used for freight transport and a freight depot still exists at Berlin-Moabit. The freight line is currently closed in the vicinities of Südkreuz and Ostkreuz.

Branches and connection curves

S-Bahn

Branches from the ring line have been built as follows:

- from Gesundbrunnen and Schönhauser Allee via Bornholmer Straße to Pankow and Schönholz (operating)
- from Treptow Park and Neukölln to Baumschulenweg (operating)
- from Jungfernheide via Wernerwerk to Gartenfeld (Siemens Railway, out of service and partially dismantled)
- from Jungfernheide via Siemensstadt-Fürstenbrunn to Spandau (S-Bahn tracks removed)

Connecting curves between the ring line and the Stadtbahn are at the Ostkreuz and Westkreuz stations.

- The south ring curve in Ostkreuz was until the evening of 28 August 2009 used by regular S-Bahn trains. It has been demolished and is being completely rebuilt. It was previously two tracks and is being rebuilt with two tracks.
- The north ring curve in Ostkreuz was closed down on 28 May 2006 and dismantled.
- A connection between Charlottenburg and Messe Nord/ICC (north ring curve) was used until 1944 and after destruction in World War II was not rebuilt.
- The connecting curve between Charlottenburg and Halensee (south ring curve) was rebuilt in the early 1990s with only one track. Currently it is used for service traffic and on weekdays by two services of line S46 each working day.

The Südringspitzkehre spur to Potsdamer Bahnhof was closed in 1944 due to war damage and never rebuilt. Its reconstruction is being considered in the planning options for line S21.

Mainline

The following long-distance and freight curves have been built connecting to the ring line:

- from Berlin-Moabit, for freight from the west, formerly connecting to the Hamburger Bahnhof and Lehrter Bahnhof, still used for freight.
- in Wedding/Westhafen since 2006, connecting to the North-South mainline from both directions towards the Hauptbahnhof
- in Gesundbrunnen/Schönhauser Allee, connecting from both directions to the Stettin Railway and the northeast
- in Frankfurter Allee/Ostkreuz, from both directions to the Berlin Frankfurter Allee–Berlin-Rummelsburg line to Berlin-Lichtenberg station and the Rummelsburg marshalling yard
- in the Treptow Park area, from the north to the Görlitz line (currently out of service)
- in Neukölln, from the west to the Görlitz line
- in Hermannstraße, from the east to the Neukölln–Mittenwald line
- in Tempelhof/Südkreuz, a freight rail from the east to Berlin-Marienfelde (currently out of service)
- in Südkreuz/Schöneberg, a freight rail towards Zehlendorf (currently out of service)
- in Westkreuz/Halensee from both directions to the Wetzlar line

- in Westend/Jungfernheide from both directions to Spandau

Notes

[1] (German) "Rekordfahrgastzahlen bei der S-Bahn" (http://www.db.de/site/bahn/de/unternehmen/presse/presseinformationen/bbmv/bbmv20061228.html). Deutsche Bahn AG. .

[2] (German) "AUS DER GESCHICHTE DER BERLINER RINGBAHN Der "Hundekopf" entsteht wieder". Berliner Zeitung. 2001-09-13. pp. p. S 07.

References

- Leo Favier, Aisha Ronniger Andrea Schulz, Alexander Schug, ed (2009) (in German). *Ring frei! Erkundungstour Ringbahn Berlin (Clear the ring! Exploring the Berlin Ring line).* Berlin: Vergangenheitsverlag. ISBN 978-3-940621-04-7.
- Berliner S-Bahn Museum (2002) (in German). *Strecke ohne Ende—Die Berliner Ringbahn (Line without end—The Berlin ring line).* Berlin: Verlag GVE. ISBN 3-892-18074-1.
- Bienert, Michael; Hoppe, Ralph (2002) (in German). *Eine Stunde Stadt (One hour city).* Berlin: Berlin Edition. ISBN 3-814-80096-6.
- Bley, Peter (1974). "50 Jahre Berliner S-Bahn (50 years of the Berlin S-Bahn)" (in German). *Berliner Verkehrsblätter* (21).
- Bley, Peter (1997) (in German). *Die Berliner S-Bahn: Gesellschaftsgeschichte eines industriellen Verkehrsmittels (The Berlin S-Bahn: social history of a transport industry).* Düsseldorf: Alba.
- Bley, Peter (1980) (in German). *Berliner S-Bahn: vom Dampfzug zur elektrischen Stadtschnellbahn (Berlin S-Bahn: from steam to electric rapid transport).* Düsseldorf: Alba,.
- Suadicani, Waldemar (1915). "Berliner Ringbahn" (http://www.zeno.org/Roell-1912/A/Berliner+Ringbahn?hl=suadicani) (in German). Berlin, Wien: Röll: Enzyklopädie des Eisenbahnwesens.

External links

- Pictures of the Ringbahn (http://www.stadtschnellbahn-berlin.de/index.php?http://www.stadtschnellbahn-berlin.de/strecken/02/index.html) (German)

Berlin Wall

The **Berlin Wall** (German: *Berliner Mauer*) was a barrier constructed by the German Democratic Republic (GDR, East Germany) starting on 13 August 1961, that completely cut off West Berlin from surrounding East Germany and from East Berlin. The barrier included guard towers placed along large concrete walls,[1] which circumscribed a wide area (later known as the "death strip") that contained anti-vehicle trenches, "fakir beds" and other defenses. The Eastern Bloc claimed that the wall was erected to protect its population from fascist elements conspiring to prevent the "will of the people" in building a socialist state in East Germany. In practice, the Wall served to prevent the massive emigration and defection that marked Germany and the communist Eastern Bloc during the post-World War II period.

View from the West Berlin side of graffiti art on the wall in 1986. The wall's infamous "death strip", on the east side of the wall, here follows the curve of the long closed Luisenstadt Canal.

The Berlin Wall was officially referred to as the "**Anti-Fascist Protection Rampart**" (German: *Antifaschistischer Schutzwall*) by GDR authorities, implying that neighbouring West Germany had not been fully de-Nazified.[2] The West Berlin city government sometimes referred to it as the "Wall of Shame"—a term coined by mayor Willy Brandt—while condemning the Wall's restriction on freedom of movement. Along with the separate and much longer Inner German border (IGB) that demarcated the border between East and West Germany, both borders came to symbolize the "Iron Curtain" that separated Western Europe and the Eastern Bloc during the Cold War.

Map of the location of the Berlin Wall, showing checkpoints

Before the Wall's erection, 3.5 million East Germans circumvented Eastern Bloc emigration restrictions and defected from the GDR, many by crossing over the border from East Berlin into West Berlin, from where they could then travel to West Germany and other Western European countries. Between 1961 and 1989, the wall prevented almost all such emigration.[3] During this period, around 5,000 people attempted to escape over the wall, with estimates of the resulting death toll varying between 100 and 200.

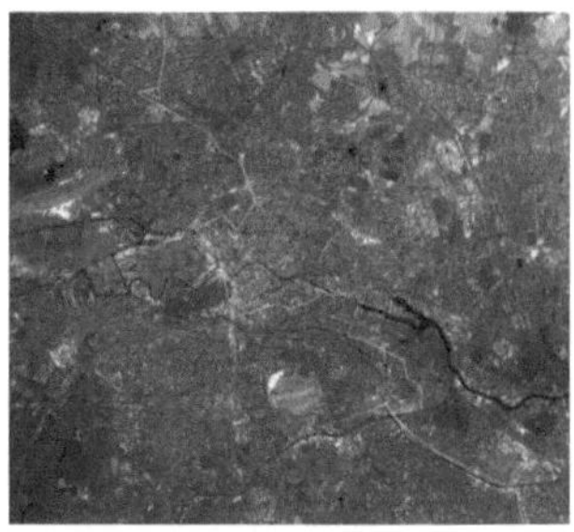

Satellite image of Berlin, with the wall's location marked in yellow

In 1989, a radical series of political changes occurred in the Eastern Bloc, associated with the liberalization of the Eastern Bloc's authoritarian systems and the erosion of political power in the pro-Soviet governments in nearby Poland and Hungary. After several weeks of civil unrest, the East German government announced on 9 November 1989 that all GDR citizens could visit West Germany and West Berlin. Crowds of East Germans crossed and climbed onto the wall, joined by West Germans on the other side in a celebratory atmosphere. Over the next few weeks, a euphoric public and souvenir hunters chipped away parts of the wall; the governments later used industrial equipment to remove most of the rest. The fall of the Berlin Wall paved the way for German reunification, which was formally concluded on 3 October 1990.

Background

Post-war Germany

After the end of World War II in Europe, what remained of pre-war Germany west of the Oder-Neisse line was divided into four occupation zones (per the Potsdam Agreement), each one controlled by one of the four occupying Allied powers: the United States, the United Kingdom, France and the Soviet Union. The capital of Berlin, as the seat of the Allied Control Council, was similarly subdivided into four sectors despite the city's location fully within the Soviet zone.[4]

Within two years, political divisions increased between the Soviets and the other occupying powers. These included the Soviets' refusal to agree to reconstruction plans making post-war Germany self-sufficient and to a detailed accounting of the industrial plants, goods and infrastructure already removed by the Soviets.[5] Britain, France, the United States and the Benelux countries later met to combine the non-Soviet zones of the country into one zone for reconstruction and approve the extension of the Marshall Plan.

The Eastern Bloc and the Berlin airlift

Further information: Eastern Bloc and Berlin Blockade

Following World War II, Soviet leader Joseph Stalin headed a union of nations on his Western border, the Eastern Bloc, that then included Poland, Hungary and Czechoslovakia, which he wished to maintain alongside a weakened Soviet-controlled Germany.[6] As early as 1945, Stalin revealed to German communist leaders that he expected to slowly undermine the British position within the British occupation zone, that the United States would withdraw within a year or two, and that nothing then would stand in the way of a united communist Germany within the bloc.[7]

The major task of the ruling communist party in the Soviet zone was to channel Soviet orders down to both the administrative apparatus and the other bloc parties, which in turn would be presented as internal measures.[8] Property and industry was nationalized in the East German zone.[9] [10] If statements or decisions deviated from the described line, reprimands and, for persons outside public attention, punishment would ensue, such as imprisonment, torture and even death.[8]

Indoctrination of Marxism-Leninism became a compulsory part of school curricula, sending professors and students fleeing to the West. The East Germans created an elaborate political police apparatus that kept the population under close surveillance,[11] including Soviet SMERSH secret police.[9]

In 1948, following disagreements regarding reconstruction and a new German currency, Stalin instituted the Berlin Blockade, preventing food, materials and supplies from arriving in West Berlin.[12] The United States, Britain, France, Canada, Australia, New Zealand and several other countries began a massive "Berlin airlift", supplying West Berlin with food and other supplies.[13] The Soviets mounted a public relations campaign against the western policy change. Communists attempted to disrupt the elections of 1948, preceding large losses therein,[14] while 300,000 Berliners demonstrated for the international airlift to continue.[15] In May 1949, Stalin lifted the blockade, permitting the resumption of Western shipments to Berlin.[16] [17]

The German Democratic Republic (East Germany) was declared on 7 October 1949. By a secret treaty, the Soviet Ministry of Foreign Affairs accorded the East German state administrative authority, but not autonomy. The Soviets penetrated East German administrative, military and secret police structures and had full control.[18] [19]

East Germany differed from West Germany (Federal Republic of Germany), which developed into a Western capitalist country with a social market economy (*"Soziale Marktwirtschaft"* in German) and a democratic parliamentary government. Continual economic growth starting in the 1950s fuelled a 20-year "economic miracle" (*"Wirtschaftswunder"*). As West Germany's economy grew and its standard of living steadily improved, many East Germans wanted to move to West Germany.

Emigration westward in the early 1950s

After Soviet occupation of Eastern Europe at the end of World War II, the majority of those living in the newly acquired areas of the Eastern Bloc aspired to independence and wanted the Soviets to leave.[20] Taking advantage of the zonal border between occupied zones in Germany, the number of GDR citizens moving to West Germany totaled 187,000 in 1950; 165,000 in 1951; 182,000 in 1952; and 331,000 in 1953.[21] [22] One reason for the sharp 1953 increase was fear of potential further Sovietization, given the increasingly paranoid actions of Joseph Stalin in late 1952 and early 1953.[23] 226,000 had fled in just the first six months of 1953.[24]

Erection of the inner German border

Further information: Inner German border and Eastern Bloc emigration and defection

By the early 1950s, the Soviet approach to controlling national movement, restricting emigration, was emulated by most of the rest of the Eastern Bloc, including East Germany.[25] The restrictions presented a quandary for some Eastern Bloc states that had been more economically advanced and open than the Soviet Union, such that crossing borders seemed more natural—especially where no prior border existed between East and West Germany.[26]

Up until 1952, the lines between East Germany and the western occupied zones could be easily crossed in most places.[27] On 1 April 1952, East German leaders met the Soviet leader Joseph Stalin in Moscow; during the discussions Stalin's foreign minister Vyacheslav Molotov proposed that the East Germans should "introduce a system of passes for visits of West Berlin residents to the territory of East Berlin [so as to stop] free movement of Western agents" in the GDR. Stalin agreed, calling the situation "intolerable". He advised the East Germans to build up their border defenses, telling them that "The demarcation line between East and West Germany should be considered a border—and not just any border, but a dangerous one ... The Germans will guard the line of defence with their lives."[28]

Consequently, the inner German border between the two German states was closed, and a barbed-wire fence erected. The border between the Western and Eastern sectors of Berlin, however, remained open, although traffic between the Soviet and the Western sectors was somewhat restricted. This resulted in Berlin becoming a magnet for East Germans desperate to escape life in the GDR, and also a flashpoint for tension between the United States and the Soviet Union.

In 1955, the Soviets gave East Germany authority over civilian movement in Berlin, passing control to a regime not recognized in the West.[29] Initially, East Germany granted "visits" to allow its residents access to West Germany. However, following the defection of large numbers of East Germans under this regime, the new East German state legally restricted virtually all travel to the West in 1956.[27] Soviet East German ambassador Mikhail Pervukhin observed that "the presence in Berlin of an open and essentially uncontrolled border between the socialist and capitalist worlds unwittingly prompts the population to make a comparison between both parts of the city, which unfortunately, does not always turn out in favor of the Democratic [East] Berlin."[30]

The Berlin emigration loophole

Further information: Eastern Bloc emigration and defection

With the closing of the inner German border officially in 1952,[30] the border in Berlin remained considerably more accessible then because it was administered by all four occupying powers.[27] Accordingly, Berlin became the main route by which East Germans left for the West.[31] On 11 December 1957, East Germany introduced a new passport law that reduced the overall number of refugees leaving Eastern Germany.

It had the unintended result of drastically increasing the percentage of those leaving through West Berlin from 60% to well over 90% by the end of 1958.[30] Those caught trying to leave East Berlin were subjected to heavy penalties, but with no physical barrier and subway train access still available to West Berlin, such measures were ineffective.[32] The Berlin sector border was essentially a "loophole" through which Eastern Bloc citizens could still escape.[30] The 3.5 million East Germans who had left by 1961 totalled approximately 20% of the entire East German population.[32]

Brain drain

Further information: Eastern Bloc emigration and defection

The emigrants tended to be young and well-educated, leading to the "brain drain" feared by officials in East Germany.[20] Yuri Andropov, then the CPSU Director on Relations with Communist and Workers Parties of Socialist Countries, wrote an urgent letter on 28 August 1958, to the Central Committee about the significant 50% increase in the number of East German intelligentsia among the refugees.[33] Andropov reported that, while the East German leadership stated that they were leaving for economic reasons, testimony from refugees indicated that the reasons were more political than material.[33] He stated "the flight of the intelligentsia has reached a particularly critical phase."[33]

By 1960, the combination of World War II and the massive emigration westward left East Germany with only 61% of its population of working age, compared to 70.5% before the war.[32] The loss was disproportionately heavy among professionals: engineers, technicians, physicians, teachers, lawyers and skilled workers.[32] The direct cost of manpower losses to East Germany (and corresponding gain to the West) has been estimated at $7 billion to $9 billion, with East German party leader Walter Ulbricht later claiming that West Germany owed him $17 billion in compensation, including reparations as well as manpower losses.[32] In addition, the drain of East Germany's young population potentially cost it over 22.5 billion marks in lost educational investment.[34] The brain drain of professionals had become so damaging to the political credibility and economic viability of East Germany that the re-securing of the German communist frontier was imperative.[35]

Construction begins, 1961

On 15 June 1961, First Secretary of the Socialist Unity Party and GDR State Council chairman Walter Ulbricht stated in an international press conference, *"Niemand hat die Absicht, eine Mauer zu errichten!"* (No one has the intention of erecting a wall!). It was the first time the colloquial term *Mauer* (wall) had been used in this context.

East German Combat Groups of the Working Class close the border on 13 August 1961 in perparation of the Berlin Wall construction.

The record of a telephone call between Nikita Khrushchev and Ulbricht on 1 August in the same year, suggests that it was Khrushchev from whom the initiative for the construction of the wall came.[36] [37] However, other sources suggest that Khrushchev had initially been wary about building a wall, fearing negative Western reaction. What is beyond dispute, though, is that Ulbricht had pushed for a border closure for quite some time, arguing that East Germany's very existence was at stake.[38] Khrushchev had been emboldened by US President John F. Kennedy's tacit indication that the US would not actively oppose this action in the Soviet sector of Berlin.[39] On Saturday, 12 August 1961, the leaders of the GDR attended a garden party at a government guesthouse in Döllnsee, in a wooded area to the north of East Berlin. There Ulbricht signed the order to close the border and erect a wall.

At midnight, the police and units of the East German army began to close the border and, by Sunday morning, 13 August, the border with West Berlin was closed. East German troops and workers had begun to tear up streets running alongside the border to make them impassable to most vehicles and to install barbed wire entanglements and fences along the 156 kilometres (97 mi) around the three western sectors, and the 43 kilometres (27 mi) that divided West and East Berlin.

East German construction workers building the Berlin Wall, 20 November 1961

The barrier was built slightly inside East Berlin or East German territory to ensure that it did not encroach on West Berlin at any point. Later, it was built up into the Wall proper, the first concrete elements and large blocks being put in place on 17 August. During the construction of the Wall, National People's Army (NVA) and Combat Groups of the Working Class (KdA) soldiers stood in front of it with orders to shoot anyone who attempted to defect. Additionally, chain fences, walls, minefields and other obstacles were installed along the length of East Germany's western border with West Germany proper. A huge no man's land was cleared to provide a clear line of fire at fleeing refugees.[40]

Immediate effects

With the closing of the East-West sector boundary in Berlin, the vast majority of East Germans could no longer travel or emigrate to West Germany. Many families were split, while East Berliners employed in the West were cut off from their jobs. West Berlin became an isolated exclave in a hostile land. West Berliners demonstrated against the wall, led by their Mayor (Oberbürgermeister) Willy Brandt, who strongly criticized the United States for failing to respond. Allied intelligence agencies had hypothesized about a wall to stop the flood of refugees, but the main candidate for its location was around the perimeter of the city. In 1961, Secretary of State Dean Rusk proclaimed, "The Wall certainly ought not to be a permanent feature of the European landscape. I see no reason why the Soviet Union should think it is—it is to their advantage in any way to leave there that monument to Communist failure."[40]

In a speech on 26 July 1963, US President John F. Kennedy had acknowledged[41] that the United States could only hope to defend West Berliners and West Germans; to attempt to stand up for East Germans would result only in an embarrassing downfall. Accordingly, the administration made polite protests at length via the usual channels, but without fervour. The Wall violated postwar Potsdam Agreements, which gave the United Kingdom, France and the United States a say over the administration of the whole of Berlin. A few months after the barbed wire was erected, the U.S. government informed the Soviet government that it accepted the Wall as "a fact of international life" and would not challenge it by force.

US President John F. Kennedy visiting the Berlin Wall on 26 June 1963

From the East-Side 1968 the "Baby Wall" with flowers was the nearest point for visitors

US and UK sources had expected the Soviet sector to be sealed off from West Berlin, but were surprised by how long the East Germans took for such a move. They considered the wall as an end to concerns about a GDR/Soviet retaking or capture of the whole of Berlin; the wall would presumably have been an unnecessary project if such plans were afloat. Thus they concluded that the possibility of a Soviet military conflict over Berlin decreased.[42]

The East German government claimed that the Wall was an "anti-fascist protective rampart" (German: *"antifaschistischer Schutzwall"*) intended to dissuade aggression from the West.[43] Another official justification was the activities of western agents in Eastern Europe.[44] The Eastern German government also claimed that West Berliners were buying out state-subsidized goods in East Berlin. East Germans and others greeted such statements with skepticism, as most of the time, the border was only closed for citizens of East Germany traveling to the West, but not for residents of West Berlin travelling to the East.[45] The construction of the Wall had caused considerable hardship to families divided by it. Most people believed that the Wall was mainly a means of preventing the citizens of East Germany from entering or fleeing to West Berlin.

An East German SED propaganda booklet published in 1955 dramatically described the serious nature of 'flight from the republic':

> Both from the moral standpoint as well as in terms of the interests of the whole German nation, leaving the GDR is an act of political and moral backwardness and depravity. Those who let themselves be recruited objectively serve West German Reaction and militarism, whether they know it or not. Is it not despicable when for the sake of a few alluring job offers or other false promises about a "guaranteed future" one leaves a country in which the seed for a new and more beautiful life is sprouting, and is already showing the first fruits, for the place that favours a new war and destruction?
>
> Is it not an act of political depravity when citizens, whether young people, workers, or members of the intelligentsia, leave and betray what our people have created through common labour in our republic to offer themselves to the American or British secret services or work for the West German factory owners, Junkers, or militarists? Does not leaving the land of progress for the morass of an historically outdated social order demonstrate political backwardness and blindness? ...
>
> [W]orkers throughout Germany will demand punishment for those who today leave the German Democratic Republic, the strong bastion of the fight for peace, to serve the deadly enemy of the German people, the imperialists and militarists.[46]

Secondary response

Kennedy appointed retired General Lucius D. Clay, who had been the Military Governor of the US Zone of Occupation in Germany during the period of the Berlin Blockade and had ordered the first measures in what became the Berlin Airlift, as his special advisor, sending him to Berlin with ambassadorial rank. Clay was immensely popular with the residents of West Berlin, and his appointment was an unambiguous sign that Kennedy would not compromise on the status of West Berlin. Clay and Vice President Lyndon B. Johnson arrived at Tempelhof Airport on the afternoon of Saturday, 19 August 1961.

They arrived in a city defended by three Allied brigades—one each from the UK, the US, and France (the Forces Françaises à Berlin). On 16 August, Kennedy had given the order for them to be reinforced. Early on 19 August, the 1st Battle Group, 18th Infantry (commanded by Colonel Glover S. Johns Jr.) was alerted.[47]

On Sunday morning, U.S. troops marched from West Germany through East Germany, bound for West Berlin. Lead elements—arranged in a column of 491 vehicles and trailers carrying 1,500 men, divided into five march units—left the Helmstedt-Marienborn checkpoint at 06:34. At Marienborn, the Soviet checkpoint next to Helmstedt on the West German/East German border, US personnel were counted by guards. The column was 160 kilometres (99 mi) long, and covered 177 kilometres (110 mi) from Marienborn to Berlin in full battle gear. East German police watched from beside trees next to the autobahn all the way along.

The front of the convoy arrived at the outskirts of Berlin just before noon, to be met by Clay and Johnson, before parading through the streets of Berlin in front of a large crowd. At 04:00 on 21 August, Lyndon Johnson left West Berlin in the hands of Gen. Frederick O. Hartel and his brigade of 4,224 officers and men. Every three months for the next three and a half years, a new American battalion was rotated into West Berlin; each traveled by autobahn to demonstrate Allied rights.

The creation of the wall had important implications for both German states. By stemming the exodus of people from East Germany, the East German government was able to reassert its control over the country: in spite of discontent with the wall, economic problems caused by dual currency and the black market were largely eliminated. The economy in the GDR began to grow. But, the wall proved a public relations disaster for the communist bloc as a whole. Western powers used it in propaganda as a symbol of communist tyranny, particularly after East German border guards shot and killed would-be defectors. Such fatalities were later treated as acts of murder by the reunified Germany.

Structure and adjacent areas

Layout and modifications

The Berlin Wall was more than 140 kilometres (87 mi) long. In June 1962, a second, parallel fence some 100 metres (110 yd) farther into East German territory was built. The houses contained between the fences were razed and the inhabitants relocated, thus establishing what later became known as the *Death Strip*. The Death Strip was covered with raked sand or gravel, rendering footprints easy to notice, easing the detection of trespassers and also enabling officers to see which guards had neglected their task;[48] it offered no cover; and most importantly, it offered clear fields of fire for the wall guards. Through the years, the Berlin Wall evolved through four versions:

Position and course of the Berlin Wall and its border control checkpoints (1989)

1. Wire fence (1961)
2. Improved wire fence (1962–1965)
3. Concrete wall (1965–1975)
4. *Grenzmauer 75* (Border Wall 75) (1975–1989)

The "fourth-generation wall", known officially as "Stützwandelement UL 12.11" (retaining wall element UL 12.11), was the final and most sophisticated version of the Wall. Begun in 1975[49] and completed about 1980,[50] it was constructed from 45,000 separate sections of reinforced concrete, each 3.6 metres (12 ft) high and 1.2 metres (3.9 ft) wide, and cost DDM 16,155,000 or about US$3,638,000.[51] The concrete provisions added to this version of the Wall were done so to prevent escapees from driving their cars through the barricades.[52] At strategic points the wall was constructed to a somewhat weaker standard so that East German and Soviet armored vehicles could break through easily in the event of war.[53]

The top of the wall was lined with a smooth pipe, intended to make it more difficult to scale. It was reinforced by mesh fencing, signal fencing, anti-vehicle trenches, barbed wire, dogs on long lines, "beds of nails" under balconies hanging over the "death strip", over 116 watchtowers,[54] and 20 bunkers. This version of the Wall is the one most commonly seen in photographs, and surviving fragments of the Wall in Berlin and elsewhere around the world are generally pieces of the fourth-generation Wall. The layout came to resemble the inner German border in most technical aspects, except the Berlin Wall had no landmines and no spring-guns.[48]

Surrounding municipalities

Besides the sector-sector boundary within Berlin itself, the wall also separated West Berlin from the present-day state of Brandenburg. The following present-day municipalities, listed in counter-clockwise direction, share a border with former West Berlin:

- **Oberhavel :** Mühlenbecker Land (partially), Glienicke/Nordbahn, Hohen Neuendorf, Hennigsdorf
- **Havelland :** Schönwalde-Glien, Falkensee, Dallgow-Döberitz
- **Potsdam** (Urban district)
- **Potsdam-Mittelmark :** Stahnsdorf, Kleinmachnow, Teltow
- **Teltow-Fläming :** Großbeeren, Blankenfelde-Mahlow
- **Dahme-Spreewald :** Schönefeld (partially)

Official crossings and usage

There were nine border crossings between East and West Berlin, which allowed visits by West Berliners, West Germans, Western foreigners and Allied personnel into East Berlin, as well as visits by GDR citizens and citizens of other socialist countries into West Berlin, provided that they held the necessary permits. Those crossings were restricted according to which nationality was allowed to use it (East Germans, West Germans, West Berliners, other countries). The most famous was the vehicle and pedestrian checkpoint at the corner of Friedrichstraße and Zimmerstraße, also known as Checkpoint Charlie, which was restricted to Allied personnel and foreigners.[55]

Several other border crossings existed between West Berlin and surrounding East Germany. These could be used for transit between West Germany and West Berlin, for visits by West Berliners into East Germany, for transit into countries neighbouring East Germany (Poland, Czechoslovakia, Denmark), and for visits by East Germans into West Berlin carrying a permit. After the 1972 agreements, new crossings were opened to allow West Berlin waste to be transported into East German dumps, as well as some crossings for access to West Berlin's exclaves (see *Steinstücken*).

Four autobahns connected West Berlin to West Germany, the most famous being the Berlin-Helmstedt autobahn, which entered East German territory between the towns of Helmstedt and Marienborn (Checkpoint Alpha), and which entered West Berlin at Dreilinden (Checkpoint Bravo for the Allied forces) in southwestern Berlin. Access to West Berlin was also possible by railway (four routes) and by boat for commercial shipping via canals and rivers.

The famous *You Are Leaving* sign at a border of the American sector

Non-German Westerners could cross the border at Friedrichstraße station in East Berlin and at Checkpoint Charlie. When the Wall was erected, Berlin's complex public transit networks, the S-Bahn and U-Bahn, were divided with it.[50] Some lines were cut in half; many stations were shut down. Three western lines traveled through brief sections of East Berlin territory, passing through eastern stations (called *Geisterbahnhöfe,* or ghost stations) without stopping. Both the eastern and western networks converged at Friedrichstraße, which became a major crossing point for those (mostly Westerners) with permission to cross.

Travel orders to go to Berlin as used by U.S. forces in the 1980s

Who could cross

West Germans and citizens of other Western countries could in general visit East Germany. Usually this involved application of a visa at an East German embassy several weeks in advance. Visas for day trips restricted to East Berlin were issued without previous application in a simplified procedure at the border crossing. However, East German authorities could refuse entry permits without stating a reason. In the 1980s, visitors from the western part of the city who wanted to visit the eastern part had to exchange at least DM 25 into East German currency at the poor exchange rate of 1:1. It was forbidden to export East German currency from the East, but money not spent could be left at the border for possible future visits. Tourists crossing from the west had to also pay for a visa, which cost DM 5; West Berliners did not have to pay this.

West Berliners initially could not visit East Berlin or East Germany at all. All crossing points were closed to them between 26 August 1961 and 17 December 1963. In 1963, negotiations between East and West resulted in a limited possibility for visits during the Christmas season that year (*Passierscheinregelung*). Similar, very limited arrangements were made in 1964, 1965 and 1966.

In 1971, with the Four Power Agreement on Berlin, agreements were reached that allowed West Berliners to apply for visas to enter East Berlin and East Germany regularly, comparable to the regulations already in force for West Germans. However, East German authorities could still refuse entry permits.

East Berliners and East Germans could at first not travel to West Berlin or West Germany at all. This regulation remained in force essentially until the fall of the wall, but over the years several exceptions to these rules were introduced, the most significant being:

- Elderly pensioners could travel to the West starting in 1965
- Visits of relatives for important family matters
- People who had to travel to the West for professional reasons (for example, artists, truck drivers, musicians, writers and so on.)

However, each visit had to be applied for individually and approval was never guaranteed. In addition, even if travel was approved, GDR travellers could exchange only a very small amount of East German Marks into Deutsche Marks (DM), thus limiting the financial resources available for them to travel to the West. This led to the West German practice of granting a small amount of DM annually (*Begrüßungsgeld*, or welcome money) to GDR citizens visiting West Germany and West Berlin to help alleviate this situation.

Citizens of other East European countries were in general subject to the same prohibition of visiting Western countries as East Germans, though the applicable exception (if any) varied from country to country.

Allied military personnel and civilian officials of the Allied forces could enter and exit East Berlin without submitting to East German passport controls, purchasing a visa or being required to exchange money. Likewise, Soviet military patrols could enter and exit West Berlin. This was a requirement of the post-war Four Powers Agreements. A particular area of concern for the Western Allies involved official dealings with East German authorities when crossing the border, since Allied policy did not recognize the authority of the GDR to regulate Allied military traffic to and from West Berlin, as well as the Allied presence within Greater Berlin, including entry into, exit from, and presence within East Berlin; the Allies held that only the Soviet Union, and not the GDR, had authority to regulate Allied personnel in such cases. For this reason, elaborate procedures were established to prevent inadvertent recognition of East German authority when engaged in travel through the GDR and when in East Berlin. Special rules applied to travel by Western Allied military personnel assigned to the Military Liaison Missions accredited to the commander of Soviet forces in East Germany, located in Potsdam.

Allied personnel were restricted by policy when travelling by land to the following routes:

Transit between West Germany and West Berlin

- **Road:** the Helmstedt-Berlin autobahn (A2) (Checkpoints Alpha and Bravo respectively). Soviet military personnel manned these checkpoints and processed Allied personnel for travel between the two points. Military personnel were required to be in uniform when traveling in this manner.
- **Rail:** Western Allied military personnel and civilian officials of the Allied forces were forbidden from using commercial train service between West Germany and West Berlin, due to the fact of GDR passport and customs controls when using them. Instead, the Allied forces operated a series of official (duty) trains that traveled between their respective duty stations in West Germany and West Berlin. When transiting the GDR, the trains would follow the route between Helmstedt and Griebnitzsee, just outside of West Berlin. In addition to persons traveling on official business, authorized personnel could also use the duty trains for personal travel on a space-available basis. The trains traveled only at night, and as with transit by car, Soviet military personnel handled the processing of duty train travelers.

Entry into and exit from East Berlin

- Checkpoint Charlie (as a pedestrian or riding in a vehicle)

As with military personnel, special procedures applied to travel by diplomatic personnel of the Western Allies accredited to their respective embassies in the GDR. This was intended to prevent inadvertent recognition of East German authority when crossing between East and West Berlin, which could jeopardize the overall Allied position governing the freedom of movement by Allied forces personnel within all Berlin.

Ordinary citizens of the Western Allied powers, not formally affiliated with the Allied forces, were authorized to use all designated transit routes through East Germany to and from West Berlin. Regarding travel to East Berlin, such persons could also use the Friedrichstraße train station to enter and exit the city, in addition to Checkpoint Charlie. In these instances, such travelers, unlike Allied personnel, had to submit to East German border controls.

Defection attempts

Further information: Deaths at the Berlin Wall

During the years of the Wall, around 5,000 people successfully defected to West Berlin. The number of people who died trying to cross the wall, or as a result of the wall's existence, has been disputed. The most vocal claims by Alexandra Hildebrandt, Director of the Checkpoint Charlie Museum and widow of the Museum's founder, estimated the death toll to be well above 200.[56] [57] A historic research group at the Center for Contemporary Historical Research (ZZF) in Potsdam has confirmed 136 deaths.[58] Prior official figures listed 98 as being killed.

NVA officer Conrad Schumann defecting to West Berlin during the wall's early days in 1961

The East German government issued shooting orders (Schießbefehl) to border guards dealing with defectors, though such orders are not the same as "shoot to kill" orders. GDR officials denied issuing the latter. In an October 1973 order later discovered by researchers, guards were instructed that people attempting to cross the wall were criminals and needed to be shot: "Do not hesitate to use your firearm, not even when the border is breached in the company of women and children, which is a tactic the traitors have often used".[59]

Early successful escapes involved people jumping the initial barbed wire or leaping out of apartment windows along the line, but these ended as the wall was fortified. East German authorities no longer permitted apartments near the wall to be occupied, and any building near the wall had its windows boarded and later bricked up. On 15 August 1961, Conrad Schumann was the first East German border guard to escape by jumping the barbed wire to West Berlin.[60] On 22 August 1961 Ida Siekmann was the first casualty at the Berlin Wall: she died after she jumped out of her third floor apartment at 48 Bernauer Strasse.[61] The first person to be shot and killed while trying to cross to West Berlin was Günter Litfin, a twenty-four year old tailor. He attempted to swim across the Spree Canal to freedom in West Germany on 24 August 1961, the same day that East German police had received shoot-to-kill orders to prevent anyone from escaping.[62]

Another dramatic escape was carried out on April 1963 by Wolfgang Engels, a 19-year-old civilian employee of the Nationale Volksarmee. Engels stole a Soviet armored personnel carrier from a base where he was deployed and drove it right into the wall. He was fired at and seriously wounded by border guards. But a West German policeman intervened, firing his weapon at the East German border guards. The policeman removed Engels from the vehicle, which had become entangled in the barbed wire.[63]

Memorial to the Victims of the Wall, with graffiti, 1982

East Germans successfully defected by a variety of methods: digging long tunnels under the wall, waiting for favorable winds and taking a hot air balloon, sliding along aerial wires, flying ultralights, and in one instance, simply driving a sports car at full speed through the basic, initial fortifications. When a metal beam was placed at checkpoints to prevent this kind of defection, up to four people (two in the front seats and possibly two in the boot) drove under the bar in a sports car that had been modified to allow the roof and windscreen to come away when it made contact with the beam. They lay flat and kept driving forward. The East Germans then built zig-zagging roads at checkpoints. The sewer system predated the wall, and some people escaped through the sewers, in a number of cases with assistance from the Girmann student group.

An airborne escape was made by Thomas Krüger, who landed a Zlin Z 42M light aircraft of the Gesellschaft für Sport und Technik, an East German youth military training organization, at RAF Gatow. His aircraft, registration DDR-WOH, was dismantled and returned to the East Germans by road, complete with humorous slogans painted on by RAF airmen such as "Wish you were here" and "Come back soon". DDR-WOH is still flying today, but under the registration D-EWOH.

If an escapee was wounded in a crossing attempt and lay on the death strip, no matter how close they were to the Western wall, Westerners could not intervene for fear of triggering engaging fire from the 'Grepos', the East Berlin border guards. The guards often let fugitives bleed to death in the middle of this ground, as in the most notorious failed attempt, that of Peter Fechter (aged 18). He was shot and bled to death, in full view of the Western media, on 17 August 1962. Fechter's death created negative publicity worldwide that led the leaders of East Berlin to place more restrictions on shooting in public places, and provide medical care for possible "would-be escapers".[64] The last person to be shot while trying to cross the border was Chris Gueffroy on 6 February 1989.

The Wall gave rise to a widespread sense of desperation and oppression in East Berlin, as expressed in the private thoughts of one resident, who confided to her diary "Our lives have lost their spirit...we can do nothing to stop them."[65]

"Mr. Gorbachev, tear down this wall."

In a speech at the Brandenburg Gate commemorating the 750th anniversary of Berlin[66] on 12 June 1987, Ronald Reagan challenged Mikhail Gorbachev, then the General Secretary of the Communist Party of the Soviet Union, to tear down the wall as a symbol of increasing freedom in the Eastern Bloc:

Ronald Reagan at the Berlin Wall, June 1987: "Tear down this wall!"

> We welcome change and openness; for we believe that freedom and security go together, that the advance of human liberty can only strengthen the cause of world peace. There is one sign the Soviets can make that would be unmistakable, that would advance dramatically the cause of freedom and peace. General Secretary Gorbachev, if you seek peace, if you seek prosperity for the Soviet Union and eastern Europe, if you seek liberalization, come here to this gate. Mr. Gorbachev, open this gate. Mr. Gorbachev, tear down this wall![67]

The Fall

Germans standing on top of the wall, 1989; it would begin to be torn apart in the following days.

After allowing for loopholes throughout the summer, Hungary effectively disabled its physical border defenses with Austria on 19 August 1989 and, in September, more than 13,000 East German tourists escaped through Hungary to Austria.[68] This set up a chain of events. The Hungarians prevented many more East Germans from crossing the border and returned them to Budapest. These East Germans flooded the West German embassy and refused to return to East Germany. The East German government responded by disallowing any further travel to Hungary, but allowed those already there to return. This triggered a similar incident in neighboring Czechoslovakia. On this occasion, the East German authorities allowed them to leave, providing that they used a train which transited East Germany on the way. This was followed by mass demonstrations within East Germany itself. (*See Monday demonstrations in East Germany.*) The longtime leader of East Germany, Erich Honecker, resigned on 18 October 1989 and was replaced by Egon Krenz a few days later. Honecker had predicted in January of that year that the wall would stand for 50 or 100 more years[69] if the conditions that had caused its construction did not change.

Berlin Wall at the Brandenburg Gate, 10 November 1989

Protest demonstrations broke out all over East Germany in September 1989. Initially, protesters were mostly people wanting to leave to the West, chanting *"Wir wollen raus!" ("We want out!"). Then protestors began to chant* "Wir bleiben hier", *("We're staying here!"). This was the start of what East Germans generally call the "Peaceful Revolution" of late 1989.*[70] *The protest demonstrations grew considerably by early November. The movement neared its height on 4 November when half a million people gathered at the Alexanderplatz demonstration, a rally for change in East Berlin's large public square and transportation hub. (Henslin, 07)*

Meanwhile, the wave of refugees leaving East Germany for the West had increased and had found its way through Hungary via Czechoslovakia (or via the West German Embassy in Prague), tolerated by the new Krenz government and in agreement with the communist Czechoslovak government. To ease the complications, the politburo led by Krenz decided on 9 November to allow refugees to exit directly through crossing points between East Germany and West Germany, including West Berlin. On the same day, the ministerial administration modified the proposal to include private travel. The new regulations were to take effect the next day.

Günter Schabowski, the party boss in East Berlin and the spokesman for the SED Politburo, had the task of announcing this; however he had not been involved in the discussions about the new regulations and had not been fully updated.[71] Shortly before a press conference on 9 November, he was handed a note announcing the changes, but given no further instructions on how to handle the information. These regulations had only been completed a few hours earlier and were to take effect the following day, so as to allow time to inform the border guards—however, nobody had informed Schabowski. He read the note out loud at the end of the conference. One of the reporters—by most accounts, NBC's Tom Brokaw--asked when the regulations would take effect. After a few seconds' hesitation, Schabowski assumed it would be the same day based on the wording of the note and replied, "As far as I know effective immediately, without delay".[38] After further questions from journalists he confirmed that the regulations included the border crossings towards West Berlin, which he had not mentioned until then.[72]

Juggling on the Berlin Wall on 16 November 1989

"Mauerspecht" (November 1989)

Walking through Checkpoint Charlie, 10 November 1989

Excerpts from Schabowski's press conference were the lead story on West Germany's two main news programs that night—at 7:17 PM on ZDF's *heute* and at 8 PM on ARD's *Tagesschau*; this of course meant that the news was broadcast to nearly all of East Germany as well. Later that night, on ARD's *Tagesthemen*, anchorman Hans Joachim Friedrichs proclaimed, "This is a historic day. East Germany has announced that, starting immediately, its borders are open to everyone. The GDR is opening its borders ... the gates in the Berlin Wall stand open."[38] [71]

After hearing the broadcast, East Germans began gathering at the wall and at the six checkpoints between East and West, demanding that border guards immediately open the gates.[71] The surprised and overwhelmed guards made many hectic telephone calls to their superiors about the problem. At first, they were ordered to find the "more aggressive" people gathered at the gates and stamp their passports with a special stamp that barred them from returning to East Germany—in effect, revoking their citizenship. However, this still left thousands of people demanding to be let through "as Schabowski said we can."[38]

It soon became clear that no one among the East German authorities would take personal responsibility for issuing orders to use lethal force, so the vastly outnumbered soldiers had no way to hold back the huge crowd of East

German citizens. Finally, at 10:45 pm, the guards finally yielded, opening the checkpoints and allowing people through with little or no identity checking. As the *Ossis* swarmed through, they were greeted by *Wessis* waiting with flowers and champagne amid wild rejoicing. Soon afterward, a crowd of West Berliners jumped on top of the wall, and were soon joined by East German youngsters. They danced together to celebrate their new freedom.

Demolition

An East German guard talks to a Westerner through a broken seam in the wall. Late November 1989.

A crane removing a section of the Berlin Wall near Brandenburg Gate on 21 December 1989

The date on which the Wall fell is considered to have been 9 November 1989 but the Wall in its entirety was not torn down immediately. Starting that evening and in the days and weeks that followed, people came to the wall with sledgehammers or otherwise hammers and chisels to chip off souvenirs, demolishing lengthy parts of it in the process and creating several unofficial border crossings. These people were nicknamed "Mauerspechte" (wall woodpeckers).

The East German regime announced the opening of ten new border crossings the following weekend, including some in historically significant locations (Potsdamer Platz, Glienicker Brücke, Bernauer Straße). Crowds on both sides waited there for hours, cheering at the bulldozers which took parts of the Wall away to reinstate old roads. Photos and television footage of these events is sometimes mislabelled "dismantling of the Wall", even though it was merely the construction of new crossings. New border crossings continued to be opened through the middle of 1990, including the Brandenburg Gate on 22 December 1989.

West Germans and West Berliners were allowed visa-free travel starting 23 December. Until then, they could only visit East Germany and East Berlin under restrictive conditions that involved application for a visa several days or weeks in advance and obligatory exchange of at least 25 DM per day of their planned stay, all of which hindered spontaneous visits. Thus, in the weeks between 9 November and 23 December, East Germans could actually travel more freely than Westerners.

West Germans curiously peer at East German border guards through a hole in the wall

Almost all of the remaining sections of Berlin Wall were rapidly chipped away. Photo December 1990

Television coverage of citizens demolishing sections of the wall on the evening of 9 November and the new border crossings opened weeks later, led some foreigners to think the Wall was torn down quickly. Technically, the Wall remained guarded for some time after 9 November, though at a decreasing intensity. In the first months, the East German military even tried to repair some of the damages done by the "wall peckers". Gradually these attempts ceased, and guards became more lax, tolerating the increasing demolitions and "unauthorized" border crossing through the holes. On 13 June 1990, the official dismantling of the Wall by the East German military began in Bernauer Straße. On 1 July, the day East Germany adopted the West German currency, all *de jure* border controls ceased, although the inter-German border had become meaningless for some time before that. The dismantling continued to be carried out by military units (after unification under the Bundeswehr) and lasted until November 1991. Only a few short sections and watchtowers were left standing as memorials.

Short section of Berlin Wall at Potsdamer Platz, March 2009

The fall of the Wall was the first step toward German reunification, which was formally concluded on 3 October 1990.

Opposition

In some European capitals at the time there was a deep anxiety over prospects for a reunified Germany. In September 1989, British Prime Minister Margaret Thatcher pleaded with Soviet President Mikhail Gorbachev not to let the Berlin Wall fall and confided that she wanted the Soviet leader to do what he could to stop it.[73] [74]

> "We do not want a united Germany. This would lead to a change to postwar borders and we cannot allow that because such a development would undermine the stability of the whole international situation and could endanger our security", Thatcher told Gorbachev.[73]

After the fall of the Berlin Wall, French President François Mitterrand warned Thatcher that a unified Germany could make more ground than Adolf Hitler ever had and that Europe would have to bear the consequences.[75]

Celebrations

On 25 December 1989, Leonard Bernstein gave a concert in Berlin celebrating the end of the Wall, including Beethoven's 9th symphony *(Ode to Joy)* with the word "Joy" *(Freude)* changed to "Freedom" *(Freiheit)* in the lyrics sung. The orchestra and choir were drawn from both East and West Germany, as well as the United Kingdom, France, the Soviet Union, and the United States.[76]

Roger Waters performed the Pink Floyd album *The Wall* just north of Potsdamer Platz on 21 July 1990, with guests including Bon Jovi, Scorpions, Bryan Adams, Sinéad O'Connor, Thomas Dolby, Joni Mitchell, Marianne Faithfull, Levon Helm, Rick Danko and Van Morrison. David Hasselhoff performed his song "Looking for Freedom", which was very popular in Germany at that time, standing on the Berlin wall.

Over the years, there has been a repeated controversial debate[77] whether 9 November would make a suitable German national holiday, often initiated by former members of political opposition in East Germany such as Werner Schulz.[78] Besides being the emotional apogee of East Germany's peaceful revolution, 9 November is also the date of the end of the Revolution of 1848 and the date of the 1918 abdication of Kaiser Wilhelm II and declaration of the Weimar Republic, the first German republic. However, 9 November is also the anniversary of the 1923 Beer Hall Putsch and the infamous *Kristallnacht* pogroms of the Nazis in 1938. Nobel Laureate Elie Wiesel criticized the first euphoria, noting that "they forgot that 9 Nov. has already entered into history—51 years earlier it marked the Kristallnacht."[79] As reunification was not official and complete until 3 October, that day was finally chosen as German Unity Day.

20th Anniversary Celebrations

On 9 November 2009, Berlin celebrated the 20th Anniversary of the Fall of the Berlin Wall with a "Festival of Freedom" with dignitaries from around the world in attendance for an evening celebration around the Brandenburg Gate. A high point was when over 1,000 colourfully designed foam domino tiles, each over 8 feet (2.4 m) tall, that were stacked along the former route of the wall in the city center were toppled in stages, converging in front of the Brandenburg Gate.[70]

A Berlin Twitter Wall [80] was set up to allow Twitter users to post messages commemorating the 20th anniversary. Masses of Chinese users have used it to protest the Great Firewall of China. Berlin Twitter Wall was quickly blocked by the Chinese authorities.[81] [82] [83]

In the United States, the German Embassy coordinated a public diplomacy campaign with the motto "Freedom Without Walls", to commemorate the 20th anniversary of the fall of the Berlin Wall. The campaign was focused on promoting awareness of the fall of the Berlin Wall among current college students. Students at over 30 universities participated in "Freedom Without Walls" events in late 2009. First place winner of the Freedom Without Walls Speaking Contest [84] Robert Cannon received a free trip to Berlin for 2010.[85]

An international project called *Mauerreise* (Journey of the Wall) took place in various countries. Twenty symbolic wall bricks were sent from Berlin starting in May 2009. Their destination: Korea, Cyprus, Yemen and other places where everyday life is characterised by division and border experience. In these places, the bricks will become a blank canvas for artists, intellectuals and young people to tackle the wall phenomenon.[86]

To commemorate the 20th Anniversary of the Fall of the Berlin Wall, Twinity reconstructed a true-to-scale section of the wall in virtual Berlin.[87] The MTV Europe Music Awards, on the 5 November, had U2 and Tokio Hotel perform songs dedicated to, and about the Berlin Wall. U2 performed at the Brandenburg Gate, and Tokio Hotel performed "World Behind My Wall".

Palestinians in the town of Kalandia, West Bank pulled down parts of the Israeli West Bank barrier, in a demonstration marking the 20th Anniversary of the Fall of the Berlin Wall.[88]

The International Spy Museum in Washington DC hosted a Trabant car rally where 20 Trabants gathered in recognition of the 20th anniversary of the fall of the Berlin Wall. Rides were raffled every half hour and a Trabant crashed through a Berlin Wall mock up. The Trabant was the East German people's car that many used to leave DDR after the collapse.

The Allied Museum in the Dahlem district of Berlin hosted a number of events to mark the Twentieth Anniversary of the Fall of the Berlin Wall. The museum held a Special Exhibition entitled "Wall Patrol – The Western Powers and the Berlin Wall 1961–1990" which focused on the daily patrols deployed by the Western powers to observe the situation along the Berlin Wall and the fortifications on the GDR border.[89] A sheet of "Americans in Berlin" Commemorative Cinderella stamps designed by T.H.E. Hill, the author of Voices Under Berlin, was presented to the Museum by David Guerra, Berlin veteran and webmaster of the site www.berlinbrigade.com. The stamps splendidly illustrate that even twenty years on, veterans of service in Berlin still regard their service there as one of the high points of their lives.[90]

Legacy

Remaining stretch of the Wall near Ostbahnhof in Friedrichshain called East Side Gallery, August 2006

Remains of the Wall adjacent to the Topography of Terror, August 2007

A memorial of over a thousand crosses and a segment of the wall for those who died trying to cross. The memorial stood for ten months in 2004 and 2005 before it was removed.

A "BERLINER MAUER 1961–1989" plaque near Checkpoint Charlie signifying where the wall stood

Little is left of the Wall at its original site, which was destroyed almost everywhere. Three long sections are still standing: an 80-metre (263 ft) piece of the first (westernmost) wall at the Topography of Terror, site of the former Gestapo headquarters, half way between Checkpoint Charlie and Potsdamer Platz; a longer section of the second (easternmost) wall along the Spree River near the Oberbaumbrücke, nicknamed East Side Gallery; and a third section that is partly reconstructed, in the north at Bernauer Straße, which was turned into a memorial in 1999. Some other isolated fragments and a few watchtowers also remain in various parts of the city.

None still accurately represents the Wall's original appearance. They are badly damaged by souvenir seekers. Fragments of the Wall were taken and some were sold around the world. Appearing both with and without certificates of authenticity, these fragments are now a staple on the online auction service eBay as well as German souvenir shops. Today, the eastern side is covered in graffiti that did not exist while the Wall was guarded by the armed soldiers of East Germany. Previously, graffiti appeared only on the western side. Along the tourist areas of the city centre, the city government has marked the location of the former wall by a row of cobblestones in the street. In most places only the "first" wall is marked, except near Potsdamer Platz where the stretch of both walls is marked, giving visitors an impression of the dimension of the barrier system.

Museum

15 years after the fall, a private museum rebuilt a 200-metre (656 ft) section close to Checkpoint Charlie, although not in the location of the original wall. They temporarily erected more than 1,000 crosses in memory of those who died attempting to flee to the West. The memorial was installed in October 2004 and demolished in July 2005.[91]

Cultural differences

For many years after reunification, people in Germany talked about cultural differences between East and West Germans (colloquially *Ossis* and *Wessis*), sometimes described as *Mauer im Kopf* (The wall in the head). A September 2004 poll found that 25 percent of West Germans and 12 percent of East Germans wished that East and West should be separated again by a "Wall".[92] A poll taken in October 2009 on the occasion of the 20th anniversary of the fall of the wall indicated, however, that only about a tenth of the population was still unhappy with the unification (8 percent in the East; 12 percent in the West). Although differences are still perceived between East and West, Germans make similar distinctions between North and South.[93]

A 2011 poll conducted by Russia's VTsIOM, found that more than half of all Russians do not know who built the Berlin Wall. Ten percent of people surveyed thought Berlin residents built it themselves. Six percent said Western powers built it and four percent thought it was a "bilateral initiative" of the Soviet Union and the West. Fifty-eight percent said they did not know who built it, with just 24 percent correctly naming the Soviet Union and its then-communist ally East Germany.[94]

Wall segments around the world

Not all segments of the wall were ground up as the wall was being torn down. Many segments have been given to various institutions around the world. They can be found, for instance in presidential and historical museums, lobbies of hotels and corporations, at universities and government buildings, and in public spaces around the world.

50th Anniversary commemoration

On 13 August 2011, Germany marked the 50th anniversary of East Germany beginning the erection of the Berlin Wall. Chancellor Angela Merkel joined with President Christian Wulff and Berlin Mayor Klaus Wowereit at the Bernauer Straße memorial park to remember lives and liberty. Speeches extolled freedom and a minute of silence at noon honored those who died trying to flee to the West. "It is our shared responsibility to keep the memory alive and to pass it on to the coming generations as a reminder to stand up for freedom and democracy to ensure that such injustice may never happen again," entreated Mayor Wowereit. "It has been shown once again: Freedom is invincible at the end. No wall can permanently withstand the desire for freedom," proclaimed President Wulff."[95] [96] [97] [98]

See also

- Brandenburg Gate
- Chapel of Reconciliation
- Checkpoint Charlie stand-off Stand-off between US and Soviet Tanks, October 1961
- Dissolution of the Soviet Union, 1991
- History of Solidarity
- List of walls
- Operation Gold
- Ostalgie
- Panmunjom, the Korean equivalent of the wall and the last standing front of the Cold War
- Paradigm shift
- Postmodernity

- Removal of Hungary's border fence
- The Berlin Wall (video game)
- The Wall – Live in Berlin, a rock opera/concert by Roger Waters

Media relating to the wall

- Some films specifically about the Berlin Wall:
  - *The Tunnel*, a NBC News Special documentary film broadcast in December 1962
  - *The Boy and the Wall* Spanish-Mexican co-production, 1965[99]
  - *The Spy Who Came in from the Cold*, 1965 – Cold War classic set on both sides of The Wall, from the book by John le Carré, directed by Martin Ritt
  - *Funeral in Berlin*, 1966 – Spy movie starring Michael Caine, directed by Guy Hamilton
  - *The Wicked Dreams of Paula Schultz*, 1968 – Cold War spy farce about an Olympic athlete who defects. Director George Marshall
  - *Berlin Tunnel 21*, 1981 – A made-for-TV movie about a former American officer leading an attempt to build a tunnel underneath The Wall as a rescue route
  - *Something To Do With The Wall*, 1991 – Documentary shot just before and after the Wall came down, by Ross McElwee and Marilyn Levine
  - *The Innocent*, 1993 – About the joint CIA/MI6 operation to build a tunnel under East Berlin in the 1950s. Directed by John Schlesinger
  - *The Tunnel*, 2001 – Dramatization of a collaborative tunnel under the wall. Film by Roland Suso Richter
- Some novels specifically about the Berlin Wall:
  - John le Carré, *The Spy Who Came in from the Cold*, 1963 – Classic Cold War spy fiction
  - Len Deighton, *Berlin Game*, 1983 – Classic Cold War spy fiction
  - T.H.E. Hill, *The Day Before the Berlin Wall: Could We Have Stopped It? – An Alternate History of Cold War Espionage* [100], 2010 – based on a legend told in Berlin in the 1970s
  - John Marks, *The Wall*, 1999 NYT Review [101] – an American spy defects to the East just hours before the Wall falls
  - Marcia Preston, *West of the Wall* (published as *Trudy's Promise* in North America), 2008 – Left behind in East Berlin, the heroine waits for news of her husband after he makes his escape over the Berlin Wall View the author's website [102]
  - Peter Schneider, *The Wall Jumper*, 1984 (German: *Der Mauerspringer*, 1982) – the Wall plays a central role in this novel set in Berlin of the 1980s
- Music related to the Berlin Wall
  - "Holidays In The Sun", a song by the English punk rock band The Sex Pistols prominently mentions the wall, specifically singer Johnny Rotten's fantasy of digging a tunnel under it.
  - *Over de muur*, a 1984 song by Dutch pop band Klein Orkest, about the differences between East and West Berlin during the period of the Berlin Wall.[103]
- Art related to the Berlin Wall
  - In the year 1982 the West-German artist ELSNER created about 500 artworks along the former border strip around West-Berlin as part of his work series "Border Injuries". On one of his actions he even tore down a large part of the wall,[104] installed a prepared foil of 3x2m in it and finished the painting there before the border soldiers on patrol could detect him. This performance was even recorded on video.[105] His actions are well-documented both in newspapers from that time and in recent scientific publications.[106]

Notes

[1] Jack Marck (http://www.americanheritage.com/articles/magazine/ah/2006/5/2006_5_70.shtml) "Over the Wall: A Once-in-a-Lifetime Experience" *American Heritage*, Oct. 2006.

[2] (http://homepage.univie.ac.at/ernst.grabovszki/Handout_Eingemauert.pdf)

[3] Monday, 20 Nov. 1989 (20 November 1989). "Freedom! – TIME" (http://www.time.com/time/magazine/article/0,9171,959058,00.html). TIME<!. . Retrieved 9 November 2009.

[4] Miller 2000, pp. 4–5

[5] Miller 2000, p. 16

[6] Miller 2000, p. 10

[7] Miller & 2000 p13

[8] Wettig 2008, p. 95-5

[9] Wettig 2008, p. 96

[10] The political process contrasted with that in western German zones occupied by Britain, France and the United States, where minister-presidents were chosen by freely elected parliamentary assemblies. (Turner, Henry Ashby. *The Two Germanies Since 1945: East and West*, Yale University Press, 1987, ISBN 0-300-03865-8, page 20)

[11] Turner 1987, p. 47

[12] Gaddis 2005, p. 33

[13] Miller 2000, pp. 65–70

[14] Turner 1987, p. 29

[15] Fritsch-Bournazel, Renata, *Confronting the German Question: Germans on the East-West Divide*, Berg Publishers, 1990, ISBN 0-85496-684-6, page 143

[16] Gaddis 2005, p. 34

[17] Miller 2000, pp. 180–81

[18] Wettig 2008, p. 179

[19] In a congratulatory telegram, Stalin emphasized that, with the creation of East Germany, the "enslavement of European countries by the global imperialists was rendered impossible." (Wettig, Gerhard, *Stalin and the Cold War in Europe*, Rowman & Littlefield, 2008, ISBN 0-7425-5542-9, page 179)

[20] Thackeray 2004, p. 188

[21] Bayerisches Staatsministerium für Arbeit und Sozialordnung, Familie und Frauen, *Statistik Spätaussiedler Dezember 2007* (http://www.stmas.bayern.de/migration/aussiedler/aussstat.pdf), p.3 (in German)

[22] Loescher 2001, p. 60

[23] Loescher 2001, p. 68

[24] Dale 2005, p. 17

[25] Dowty 1989, p. 114

[26] Dowty 1989, p. 116

[27] Dowty 1989, p. 121

[28] Harrison 2003, p. 240-fn

[29] Harrison 2003, p. 98

[30] Harrison 2003, p. 99

[31] Paul Maddrell, *Spying on Science: Western Intelligence in Divided Germany 1945–1961*, p. 56. Oxford University Press, 2006

[32] Dowty 1989, p. 122

[33] Harrison 2003, p. 100

[34] Volker Rolf Berghahn, *Modern Germany: Society, Economy and Politics in the Twentieth Century*, p. 227. Cambridge University Press, 1987

[35] Pearson 1998, p. 75

[36] Wiegrefe, Klaus. "Wir lassen euch jetzt ein, zwei Wochen Zeit" (http://einestages.spiegel.de/static/topicalbumbackground/4246/_wir_lassen_euch_jetzt_ein_zwei_wochen_zeit.html), *Spiegel Online – einestages*, May 2009

[37] "Record of the conversation" (http://www.welt.de/politik/article3828831/Das-Gespraech-zwischen-Ulbricht-und-Chruschtschow.html) (in (German)). Welt.de. 2009-05-30. . Retrieved 2011-08-06.

[38] Sebetsyen, Victor (2009). *Revolution 1989: The Fall of the Soviet Empire*. New York City: Pantheon Books. ISBN 0375425322.

[39] Kempe, Frederick (2011). *Berlin 1961*. Penguin Group (USA). p. 247. ISBN 0399157298.

[40] "Wall Goes Up in Berlin - Events of 1961 - Year in Review" (http://www.upi.com/Audio/Year_in_Review/Events-of-1961/Wall-Goes-Up-in-Berlin/12295509433760-7/). UPI.com. 1998-05-29. . Retrieved 2011-08-06.

[41] US Politics Guide (http://uspoliticsguide.com/US-Politics-Directory/US-Politcal-Historical-Speeches/JFK-Berlin-Wall.htm) – speech transcript

[42] Taylor, Frederick. *The Berlin Wall: 13 August 1961 – 9 November 1989*. Bloomsbury 2006

[43] "Goethe-Institut – Topics – German-German History Goethe-Institut" (http://web.archive.org/web/20080409133940/http://www.goethe.de/ges/ztg/thm/ddg/en1748571.htm). Web.archive.org. 2008-04-09. Archived from the original (http://www.goethe.de/ges/ztg/

thm/ddg/en1748571.htm) on 2008-04-09. . Retrieved 2011-08-06.

[44] *"Die Regierungen der Warschauer Vertragsstaaten wenden sich an die Volkskammer und an die Regierung der DDR mit dem Vorschlag, an der Westberliner Grenze eine solche Ordnung einzuführen, durch die der Wühltätigkeit gegen die Länder des sozialistischen Lagers zuverlässig der Weg verlegt und ringsum das ganze Gebiet West-Berlins eine verlässliche Bewachung gewährleistet wird."* Die Welt: Berlin wird geteilt (http://www.welt.de/print-welt/article252861/Berlin_wird_geteilt.html)

[45] "Neues Deutschland: Normales Leben in Berlin,14 August 1961" (http://www.zlb.de/projekte/millennium/original_html/nd_1961_1408.GIF.html) (in German). Zlb.de. . Retrieved 2011-08-06.

[46] English translation of " Wer die Deutsche Demokratische Republik verläßt, stellt sich auf die Seite der Kriegstreiber (http://www.calvin.edu/academic/cas/gpa/notiz3.htm)" ("He Who Leaves the German Democratic Republic Joins the Warmongers", *Notizbuch des Agitators* ("Agitator's Notebook"), Berlin: Socialist Unity Party's Agitation Department, November 1955.

[47] See also Hackworth, About Face

[48] According to Hagen Koch, former Stasi-officer, in Geert Mak's documentary In Europa (series), episode *1961 – DDR* (http://www.ineuropa.nl/programmas/36788896/afleveringen/40824233/), 25 January 2009

[49] Heiko Burkhardt. "Facts of Berlin Wall – History of Berlin Wall" (http://www.dailysoft.com/berlinwall/history/facts.htm). Dailysoft.com. . Retrieved 9 November 2009.

[50] P. Dousset; A. Souquet; S. Lelarge. "Berlin Wall" (http://web.archive.org/web/20060913215115/http://www.wall-berlin.org/gb/mur.htm). Web.archive.org. Archived from the original (http://www.wall-berlin.org/gb/mur.htm) on 13 September 2006. . Retrieved 9 November 2009.

[51] Heiko Burkhardt. "Fourth Generation of Berlin Wall – History of Berlin Wall" (http://www.dailysoft.com/berlinwall/history/facts_02.htm). Dailysoft.com. . Retrieved 9 November 2009.

[52] Rise and Fall of the Berlin Wall. History Channel, 2009. DVD-ROM.

[53] "Rise and Fall of the Berlin Wall". *History Channel*, 2009. DVD-ROM.

[54] Popiolek. "The Berlin wall : History of Berlin Wall : Facts" (http://web.archive.org/web/20071012011801/http://die-berliner-mauer.de/en/fakten.html). Die-berliner-mauer.de. Archived from the original (http://www.die-berliner-mauer.de/en/fakten.html) on 12 October 2007. . Retrieved 9 November 2009.

[55] Harrison 2003, pp. 206–14

[56] Chronik der Mauer: Todesopfer an der Berliner Mauer (http://www.chronik-der-mauer.de/index.php/de/Start/Index/id/593792) (in German)

[57] "Forschungsprojekt "Die Todesopfer an der Berliner Mauer, 1961–1989": BILANZ (Stand: 7. August 2008) (in German)" (http://www.chronik-der-mauer.de/index.php/de/Common/Document/field/file/id/45997). . Retrieved 2011-08-06.

[58] "Center for Contemporary Historical Research (Zentrum für Zeithistorische Forschung Potsdam e.V) in German" (http://www.chronik-der-mauer.de/index.php/de/Start/Index/id/593792). Chronik-der-mauer.de. . Retrieved 2011-08-06.

[59] "E German 'licence to kill' found" (http://news.bbc.co.uk/1/hi/world/europe/6943093.stm). BBC. 12 August 2007. . Retrieved 12 August 2007. "A newly discovered order is the firmest evidence yet that the communist regime gave explicit shoot-to-kill orders, says Germany's director of Stasi files."

[60] "Conrad Schumann, 56, Symbol of E. Berlin escapes"; North Sports Final Edition Associated Press. *Chicago Tribune*, Chicago, Ill.: 23 Jun 1998. pg. 8

[61] "Chronik der Mauer - Bau und Fall der Berliner Mauer | Opfer der Mauer" (http://www.chronik-der-mauer.de/index.php/de/Start/Detail/id/593816/page/1). Chronik-der-mauer.de. . Retrieved 2011-08-06.

[62] Kempe, Frederick (2011). *Berlin 1961*. Penguin Group (USA). pp. 363–367. ISBN 0399157298.

[63] Hertle, Hans-Hermann (2008). *The Berlin Wall: Monument of the Cold War*. Ch. Links Verlag, p. 72. ISBN 3-86153-463-0

[64] Taylor, Frederick . *The Berlin Wall: A World Divided 1961–1989*, London: Harper Perennial, 2006.

[65] Kempe, Frederick (2011). *Berlin 1961*. Penguin Group (USA). p. 394. ISBN 0399157298.

[66] "Reagan's 'tear down this wall' speech turns 20 – USATODAY.com" (http://www.usatoday.com/news/world/2007-06-12-reagan-speech_N.htm). *USA Today*. 12 June 2007. . Retrieved 19 February 2008.

[67] "Remarks at the Brandenberg Gate" (http://web.archive.org/web/20080622043537/http://www.reaganfoundation.org/reagan/speeches/wall.asp). Ronald Reagan Presidential Foundation. Archived from the original (http://www.reaganfoundation.org/reagan/speeches/wall.asp) on 22 June 2008. . Retrieved 9 February 2008.

[68] Meyer, Michael (13 September 2009). "The picnic that brought down the Berlin Wall" (http://articles.latimes.com/2009/sep/13/opinion/oe-meyer13). *LA Times*. . Retrieved 14 June 2010.

[69] "Sie wird noch in 50 Jahren stehen" (http://www.dpa.de/Detailansicht-80er-Jahre.543+M517a9d5ff09.0.html). dpa. 22 May 2009. .

[70] unknown (2009). "20 Jahre Mauerfall[[Category:Articles containing German language text (http://www.mauerfall09.de)]"]. Kulturprojekte Berlin GmbH. . Retrieved 9 April 2009.

[71] Sarotte, Mary Elise (1 Nov 2009) "How it went down: The little accident that toppled history" (http://www.washingtonpost.com/wp-dyn/content/article/2009/10/30/AR2009103001846.html) Washington Post. Retrieved 2 November 2009.

[72] Walker, Marcus (21 Oct 2009) "Did Brinkmannship Fell Berlin's Wall? Brinkmann Says It Did (http://online.wsj.com/article/SB125597721400194603.html)" The Wall Street Journal.

[73] "How Margaret Thatcher pleaded with Gorbachev not to let the Berlin Wall fall out of london" (http://www.hindu.com/2009/09/15/stories/2009091553501100.htm). *Hasan Suroor*. Hindu. 15 September 2009. . Retrieved 8 November 2009.

[74] "Thatcher told Gorbachev Britain did not want German reunification" (http://www.timesonline.co.uk/tol/news/politics/article6829735.ece). *Times* (UK). 11 September 2009. . Retrieved 8 November 2009.
[75] "United Germany might allow another Hitler, Mitterrand told Thatcher" (http://www.timesonline.co.uk/tol/news/world/europe/article6828556.ece). *Times* (UK). 10 September 2009. . Retrieved 9 November 2009.
[76] Naxos (2006). "Ode To Freedom – Beethoven: Symphony No. 9 (NTSC)" (http://www.naxos.com/catalogue/item.asp?item_code=2072038). *Naxos.com Classical Music Catalogue*. . Retrieved 26 November 2006. This is the publisher's catalogue entry for a DVD of Bernstein's Christmas 1989 "Ode to Freedom" concert. David Hasselhoff sang during the fall of the Berlin wall.
[77] Sven Felix Kellerhof, Alan Posener (2007). "Soll der 9. November Nationalfeiertag werden?" (http://debatte.welt.de/kontroverse/48213/soll+der+9+november+nationalfeiertag+werden). Welt Online. . Retrieved 22 February 2009.
[78] Jörg Aberger (7 September 2004). "Debatte: Thierse fordert neuen Nationalfeiertag" (http://www.spiegel.de/kultur/gesellschaft/0,1518,316977,00.html). *Der Spiegel*. . Retrieved 22 February 2009.
[79] *New York Times*, 17 November 1989 – Elie Wiesel, "Op-Ed in response to the fall of the wall"
[80] http://www.berlintwitterwall.com/
[81] RSF.org (http://www.rsf.org/spip.php?page=article&id_article=34897)
[82] "Mr. Hu, tear down this firewall!" (http://www.theglobeandmail.com/blogs/points-east/mr-hu-tear-down-this-firewall/article1338387/). The Globe and Mail. . Retrieved 2011-08-06.
[83] MontrealGazette.com (http://www.montrealgazette.com/technology/BERLIN+WALL+CHINA+MARKS+DATE/2193355/story.html)
[84] http://www.germany.info/Vertretung/usa/en/09__Press__InFocus__Interviews/03__Infocus/04__Without__Walls/FWW__Speaking__Winners__VA.html
[85] unknown (2009). "Freedom Without Walls" (http://www.germany.info/withoutwalls). Embassy of the Federal Republic of Germany, Washington, DC. . Retrieved 9 April 2009.
[86] "The Wall in the World 2009 – 20th anniversary of the Fall of the Wall – Goethe-Institut" (http://www.goethe.de/ges/prj/mar/enindex.htm). Goethe.de. . Retrieved 9 November 2009.
[87] "The Berlin Wall in Twinity" (http://www.twinity.com/en/wall). Twinity. 2009. . Retrieved 9 November 2009.
[88] "Palestinians break Israel's wall - Middle East" (http://english.aljazeera.net/news/middleeast/2009/11/20091191516393254 64.html). Al Jazeera English. 2009-11-09. . Retrieved 2011-08-06.
[89] Allied Museum (2009). "Wall Patrol – The Western Powers and the Berlin Wall 1961–1990 Special Exhibition" (http://www.alliiertenmuseum.de/en/4_2.php?year=&activity_id=214). Allied Museum, Berlin, Germany. . Retrieved 23 July 2010.
[90] Allied Museum (2009). "American Commemorative Stamps Presented to the Museum" (http://www.alliiertenmuseum.de/en/4_3.php?year=2009&activity_id=230). Allied Museum, Berlin, Germany. . Retrieved 23 July 2010.[]
[91] Furlong, Ray (5 July 2005). "Berlin Wall memorial is torn down" (http://news.bbc.co.uk/2/hi/europe/4651823.stm). BBC News. . Retrieved 23 February 2006.
[92] Reuters (8 September 2004). "One in 5 Germans wants Berlin Wall rebuilt" (http://web.archive.org/web/20060613190725/http://www.msnbc.msn.com/id/5942091/). MSNBC. Archived from the original (http://www.msnbc.msn.com/id/5942091/) on 13 June 2006. . Retrieved 23 February 2006.
[93] ZDF "Wochenjournal" (5 November 2009). "Große Zustimmung zur Wiedervereinigung" (http://wochenjournal.zdf.de/ZDFde/inhalt/11/0,1872,7925355,00.html). ZDF. . Retrieved 6 November 2006.
[94] "Who built Berlin Wall? Most Russians don't know | Reuters" (http://uk.reuters.com/article/idUKTRE5A44NS20091105?feedType=nl&feedName=ukoddlyenough). Uk.reuters.com. . Retrieved 2011-08-06.
[95] "Germany marks 50th anniversary of Berlin Wall" (http://www.telegraph.co.uk/news/worldnews/europe/germany/8699626/Germany-marks-50th-anniversary-of-Berlin-Wall.html). UK Telegraph. 2011-08-13. . Retrieved 2011-08-13.
[96] "Germany Marks Construction of the Berlin Wall" (http://www.foxnews.com/world/2011/08/13/germany-marks-construction-berlin-wall-1271894147/). AP. 2011-08-13. . Retrieved 2011-08-13.
[97] "Reflecting on the Berlin Wall, 50 Years After Its Construction" (http://www.history.com/news/2011/08/11/reflecting-on-the-berlin-wall-50-years-after-its-construction/). History.com. 2011-08-11. . Retrieved 2011-08-13.
[98] "50 Years Berlin Wall 1961–1989 [video clips (http://www.germanworldonline.com/index.php/50-years-berlin-wall-1961-1989/)"]. German World. 2011-08-13. . Retrieved 2011-08-13.
[99] IMDb.com (http://www.imdb.com/title/tt0058412/)
[100] http://voicesunderberlin.com/DayBefore.html
[101] http://www.nytimes.com/books/98/09/20/reviews/980920.20norchit.html
[102] http://www.marciapreston.com/index.php?option=com_content&view=article&id=3&Itemid=2
[103] "Klein Orkest biography" (http://www.sweetslyrics.com/bio-Klein Orkest.html). Sweetslyrics.com. . Retrieved 2011-08-06.
[104] "Die Mauer als Muse" (http://www.tagesspiegel.de/berlin/die-mauer-als-muse/821970.html). *The Wall fell on July 18th 1982*. Lars von Törne in Der Tagesspiegel. . Retrieved 13 March 2007.
[105] "ELSNER's Border Injuries" (http://www.galerie-son.com/artists/elsner/elsner_works.html). *artworks & video documentation*. galerie son. .
[106] "ELSNER" (http://www.galerie-son.com/artists/elsner/elsner_press.html). *press archive*. galerie son. .

References

- Böcker, Anita (1998). *Regulation of Migration: International Experiences.* Het Spinhuis. ISBN 9055890952
- Buckley, William F., Jr. (2004). *The Fall of the Berlin Wall.* Hoboken, New Jersey: John Wiley and Sons. ISBN 0-471-26736-8
- Cate, Curtis (1978). *The Ides of August: The Berlin Wall Crisis—1961.* New York City: M. Evans
- Childs, David (2001) *The Fall of the GDR: Germany's Road To Unity*, Longman, Pearsoned.co.uk (http://www.pearsoned.co.uk/Bookshop/detail.asp?item=100000000004463) 2001. ISBN 9780582315693, ISBN 0582315697
- Childs, David, *The GDR: Moscow's German Ally*, (Second Edition 1988, First Edition 1983, George Allen & Unwin, London) ISBN 0-04-354029-5, 9780043540299.
- Childs, David, (2001) *The Fall of the GDR*, Longman. ISBN 0-582-31569-7 Amazon.co.uk (http://www.amazon.co.uk/Fall-Themes-Modern-German-History/dp/0582315697)
- Childs, David, (2000) *The Two Red Flags: European Social Democracy & Soviet Communism Since 1945*, Routledge. Informaworld.com (http://www.informaworld.com/smpp/title~db=all~content=t732627883)
- Childs, David, (1991) *Germany in the Twentieth Century*, (From pre-1918 to the restoration of German unity), Batsford, Third edition. ISBN 0713467959
- Childs, David, (1987) *East Germany to the 1990s Can It Resist Glasnost?* (http://openlibrary.org/b/OL16520921M/East_Germany_to_the_1990s), The Economist Intelligence Unit. ISBN 0-85058-245-8, 9780850582451. Worldcat.org (http://www.worldcat.org/search?q=au:"Childs,+David,"&qt=hot_author)
- Dale, Gareth (2005). *Popular Protest in East Germany, 1945–1989: Judgements on the Street.* Routledge. ISBN 0714654086
- Dowty, Alan (1989). *Closed Borders: The Contemporary Assault on Freedom of Movement.* Yale University Press. ISBN 0300044984
- Gaddis, John Lewis (2005). *The Cold War: A New History.* Penguin Press. ISBN 1594200629
- Harrison, Hope Millard (2003). *Driving the Soviets Up the Wall: Soviet-East German Relations, 1953–1961.* Princeton University Press. ISBN 0691096783.
- Catudal, Honoré M. (1980). *Kennedy and the Berlin Wall Crisis.* West Berlin: Berlin Verlag
- Hertle, Hans-Hermann (2007). *The Berlin Wall.* Bonn: Federal Centre for Political Education
- Kennedy, John F.. "July 25, 1961 speech" (http://www.jfklibrary.org/Historical+Resources/Archives/Reference+Desk/Speeches/JFK/003POF03BerlinCrisis07251961.htm). Retrieved 20 October 2009.
- Loescher, Gil (2001). *The UNHCR and World Politics: A Perilous Path.* Oxford University Press. ISBN 0198297165
- Maclean, Rory (1992). *Stalin's Nose: Across the Face of Europe.* London: HarperCollins.
- Miller, Roger Gene (2000). *To Save a City: The Berlin Airlift, 1948–1949.* Texas A&M University Press. ISBN 0890969671
- Mynz, Rainer (1995). *Where Did They All Come From? Typology and Geography of European Mass Migration In the Twentieth Century; European Population Conference Congress European de Démographie.* United Nations Population Division
- Pearson, Raymond (1998). *The Rise and Fall of the Soviet Empire.* Macmillan. ISBN 0312174071
- Schneider, Peter (2005). *The Wall Jumper.* London: Penguin Classics
- Taylor, Frederick. The Berlin Wall: 13 August 1961 – 9 November 1989. Bloomsbury 2006
- Thackeray, Frank W. (2004). *Events that changed Germany.* Greenwood Publishing Group. ISBN 0313328145
- Friedrich, Thomas (writer),and Harry Hampel (photos) (1996). *Wo die Mauer War/Where was the Wall?.* Berlin: Nicolai. ISBN 3875846958
- Turner, Henry Ashby (1987). *The Two Germanies Since 1945: East and West.* Yale University Press. ISBN 0300038658
- Wettig, Gerhard (2008). *Stalin and the Cold War in Europe.* Rowman & Littlefield. ISBN 0742555429

• Luftbildatlas. Entlang der Berliner Mauer. Karten, Pläne und Fotos. Hans Wolfgang Hoffmann / Philipp Meuser (eds.) Berlin 2009. ISBN 978-3-938666-84-5

External links

- Scholars Reflect on the 20th Anniversary of the Fall of the Berlin Wall (http://humanexperience.stanford.edu/berlinwall)
- Explore the former inner German border with DW's animated video (http://www.dw-world.de/popups/popup_single_mediaplayer/0,,4418618_type_video_struct_1432_contentId_4440098,00.html)
- Berlin Wall: 20 years on (http://www.guardian.co.uk/berlin-wall-20-years-on) video series from *The Guardian*
- The CWIHP Collection on the Rise and Fall of the Berlin Wall (http://www.wilsoncenter.org/index.cfm?topic_id=1409&fuseaction=topics.categoryview&categoryid=BE949EBA-AC3C-D7AF-5B98808A1EBD6B9E)
- (English) Freedom Without Walls: German Missions in the United States (http://www.germany.info/Vertretung/usa/en/10__Press__Facts/03__Infocus/04__Without__Walls/__Main__S.html) Looking Back at the Fall of the Berlin Wall – official homepage
- Chronicle of the Wall (http://www.chronik-der-mauer.de/index.php/de/Start/Index/id/652147/) Most comprehensive multi-media source of information on this topic
- The Berlin Wall (http://www.timesonline.co.uk/tol/system/topicRoot/The_Berlin_Wall/) Original reports and pictures from The Times
- The Berlin Wall (http://www.berlin.de/mauer/index.en.html) Official Website Berlin.de
- (German) Chronik der Mauer (http://www.chronik-der-mauer.de/index.php) Chronicle of the Wall
- (German) (English) Information Berlin Wall and East-Berlin (http://www.innerdeutsche-grenze.info/)
- Bernauer Straße Memorial website (http://www.berliner-mauer-gedenkstaette.de/en/)
- (German) Information on the East German border system (http://www.passkontrolle-ddr.de/)
- Escape from GDR, a five-part documentary on YouTube on tunnelling under the Berlin (http://www.youtube.com/watch?v=O2GRQP7K0iY&feature=related)
- Allied Forces in Berlin (FR, UK & US Berlin Brigade) (http://www.western-allies-berlin.com/)
- Photographs of time of the Fall as well as updates on the current situation in Germany (http://www.guardian.co.uk/wall/)
- Reports on reinforcements to Berlin Brigade (http://www.theberlinobserver.com/1stBG18thinf.html)
- JFK speech clarifying limits of American protection (http://web.archive.org/web/20060219134229/http://www.jfklibrary.org/jfk_berlin_crisis_speech.html)
- "Berlin 1969" includes links to other sources. (http://www.berlin1969.com/)
- Berlin Life: A concise but thorough history of the wall (http://www.berlin-life.com/berlin/wall)
- Berlin Wall: Past and Present (http://berlinwall.soaringdevelopment.com/)
- Important Berlin Wall Dates (http://www.aish.com/societyWork/society/When_the_Berlin_Wall_Came_Tumbling_Down.asp)
- The Lost Border: Photographs of the Iron Curtain (http://www.brianrose.com/lostborder.htm)
- Dossier: The Fall of the Wall – New Perspectives on 1989 (http://www.goethe.de/ges/pok/dos/dos/mau/enindex.htm)
- Freedom Without Walls, 20th Anniversary Public Diplomacy Campaign by the German Embassy Washington (http://www.germany.info/withoutwalls)
- The History Channel:Berlin Wall (http://www.history.com/topics/berlin-wall)
- 'More than 1,000 died' trying to flee East Germany (http://www.telegraph.co.uk/news/worldnews/europe/germany/1438720/More-than-1000-died-trying-to-flee-East-Germany.html). The Daily Telegraph, 13 August 2003.

Images and personal accounts

- The Berlin Wall: Rise and Fall (http://www.life.com/image/first/in-gallery/36002/the-berlin-wall-rise-and-fall) – slideshow by *Life magazine*
- Berlin Wall newsreel archives (http://www.britishpathe.com/workspace.php?id=1281&display=list) at British Pathe Adobe Flash video)
- Fall of the Wall: My School Trip (http://www.thefirstpost.co.uk/55774,in-pictures,news-in-pictures,in-pictures-photographs-the-berlin-wall-my-part-in-its-fall-by-roland-pietsch-germany) – slideshow by *The First Post*
- Moments in Time 1989/1990 (http://www.wir-waren-so-frei.de/) Internet Archive presenting personal films and photos of the time of the Fall of the Berlin Wall and the German reunification, mostly licenced under CC
- Berlin, Cyber City (http://netzspannung.org/database/113713/en) Mixed Reality Installation realized by media artists M. Fleischmann & W. Strauss (http://fleischmann-strauss.de/). Video (https://www.youtube.com/watch?v=BYmnVTvGssg) on YouTube
- (German)[[Category:Articles with German language external links (http://www.chronik-der-mauer.de/index.php/de/Start/Index/id/659901)] Comprehensive Gallery (1961 to 1990) from the website Chronicle of the Wall]
- Gallery of annotated photographs of the Berlin Wall (http://galleries.fototagger.com/link.php?action=detail&id=546)
- (French)[[Category:Articles with French language external links (http://rosset.org/photography/berlin/berlin.htm)] Photos of the Berlin Wall by Georges Rosset]
- Photos of the Berlin Wall 1989 to 1999 (http://www.berlinphotos.co.uk/html/the_wall.html)
- Gallery at berlin-wall.net (http://www.berlin-wall.net/photogallery.htm)
- Berlin Wall (http://www.panorama-cities.net/berlin/berlin_wall.html) Panorama of the East Side Gallery
- One Day In Berlin: Tracing The Wall (http://czechdaily.wordpress.com/2007/09/11/one-saturday-in-berlin/)
- Berlin Wall Online (http://www.dailysoft.com/berlinwall/), Chronicle of the Berlin Wall history includes an archive of photographs and texts
- Personal Account of the Fall of the Berlin Wall (http://www.andreas.com/berlin.html)
- Berlin Wall, Past and Present (http://berlinwall.soaringdevelopment.com/), Descriptions, Videos, Images of Berlin Wall
- Personal Accounts of 9 November 1989 (http://germany.info/wallstories)
- (German)[[Category:Articles with German language external links (http://staff-www.uni-marburg.de/~nail/mauer.htm)] Photos of the Berlin Wall 1962–1990]
- Large collection of images in Berlin Wall group (http://www.flickr.com/groups/52809934@N00/) at Flickr
- Berlin hard jump for Freedom (music dedicated to Berlin Wall History (http://www.youtube.com/watch?v=9d7N2v0E0l4) at YouTube
- Images of the "Americans in Berlin" Cinderella stamps (http://voicesunderberlin.com/Stamps1.html) at the author's website.

West Berlin

Berlin (West) / Westberlin**West Berlin**	
Allied-occupied sectors of Berlin	
← 1949–1990  →  →	
 Flag	
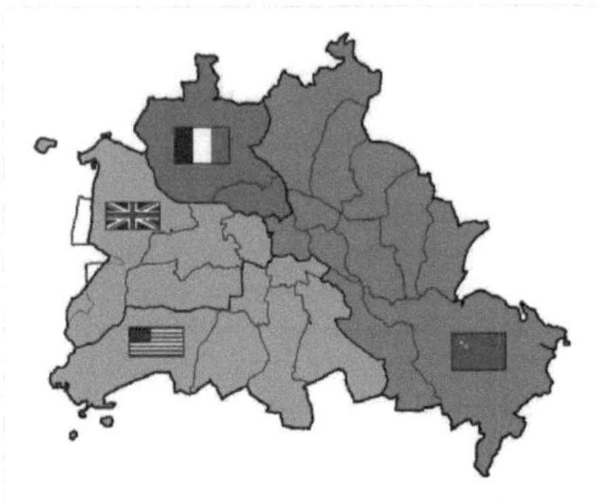 The four occupation sectors of Berlin. West Berlin is in light blue, dark blue and purple. Borough borders as of 1987.	
Historical era	Cold War
- Established	1949
- Reunification	3 October 1990

West Berlin was a political exclave that existed between 1949 and 1990. It comprised the western regions of Berlin, which were bordered by East Berlin and parts of East Germany. West Berlin consisted of the American, British, and French occupation sectors, which had been established in 1945. It was politically closely affiliated with West Germany, although it had a special status, because its administration was formally conducted by the Western Allies. East Berlin encompassed the region occupied and administered by the Soviet Union, and was claimed as capital by East Germany. The Western Allies did not recognise this claim, as they asserted that the entire city of Berlin was legally under four-power administration. The Berlin Wall, built in 1961, physically divided East and West Berlin until it fell in 1989.

With about two million inhabitants, West Berlin had the highest number of residents of any city in Cold War-era Germany.

Origins

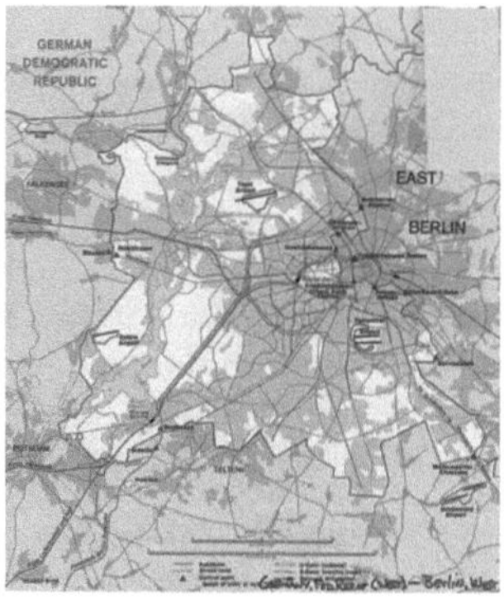

West Berlin, as of 1978

The Potsdam Agreement established the legal framework for the occupation of Germany in the wake of World War II. According to the agreement, Germany would be formally under the administration of the four major wartime Allies—the United States, the United Kingdom, France, and the Soviet Union—until a German government acceptable to all parties would be established. The territory of Germany, as it existed in 1937, would be reduced by most of *Eastern Germany* thus creating the former eastern territories of Germany. The remaining territory would be divided into four zones, each administered by one of the allied countries. Berlin, which was surrounded by the Soviet zone of occupation—newly established in most of Middle Germany—would be similarly divided, with the Western Allies occupying an enclave consisting of the western parts of the city. According to the agreement, the occupation of Berlin would end only as a result of a quadripartite agreement. The Western Allies were guaranteed three air corridors to their sectors of Berlin, and the Soviets also informally allowed road and rail access between West Berlin and the western parts of Germany (see section on traffic).

At first, this arrangement was officially only meant to be a temporary administrative structure, with all parties declaring that Germany and Berlin would soon be reunited. However, as the relations between the western allies and the Soviet Union soured and the Cold War began, the joint administration of Germany and Berlin broke down. Soon Soviet-occupied Berlin and western-occupied Berlin had separate city administrations. In 1948, the Soviets tried to force the Western Allies out of Berlin by imposing a land blockade on the western sectors: the Berlin Blockade. The West responded by using its air corridors for supplying their part of the city with food and other goods in the Berlin Airlift. In May 1949, the Soviets lifted the blockade, and West Berlin as a separate city with its own jurisdiction was maintained. By the end of 1949, two new states had been created out of occupied Germany—the *Federal Republic of Germany* (West Germany) in the West and the *German Democratic Republic* (GDR, East Germany) in the East—with West Berlin an enclave surrounded by, but not part of, the GDR.

Legal status

According to the legal theory followed by the Western Allies, the occupation of most of Germany ended in 1949 with the declaration of the Federal Republic of Germany (23 May 1949) and the German Democratic Republic (7 October 1949). However, because the occupation of Berlin could only be ended by a quadripartite agreement, Berlin remained an occupied territory under the formal sovereignty of the allies. Hence, the *Grundgesetz* (the constitution of the Federal Republic) was not applicable to West Berlin. In addition, West German federal law did not apply to West Berlin, but the House of Representatives of Berlin (German: *Abgeordnetenhaus von Berlin*; the West Berlin legislature; the reunited Berlin's legislature bears the same name) used to vote in every new federal law without debate to maintain legal status with the pre-1990 Federal Republic of Germany.

In 1969 U.S. military vehicles roar through rush hour traffic in the residential district of Zehlendorf, a routine reminder that West Berlin was still legally occupied by the World War II Allies.

The Western Allies remained the ultimate political authorities in West Berlin. All legislation of the "Abgeordnetenhaus", the domestic state and the adopted federal law, only applied under the proviso of the confirmation by the three Western Allied commanders-in-chief. If they approved a bill, it was enacted as part of West Berlin's statutory law. If the commanders-in-chief rejected a bill, it did not become law in West Berlin; this, for example, was the case with West German laws on military duty. West Berlin was run by the elected 'Governing Mayor' and the Senate of Berlin (city government) seated at Rathaus Schöneberg. Governing Mayor and Senators (ministers) had to be approved by the Western Allies and thus derived their authority from the occupying forces, not from their electoral mandate.

The Soviets unilaterally declared the occupation of East Berlin at an end along with the rest of East Germany. This move was, however, not recognised by the Western Allies, who continued to view all of Berlin as a jointly occupied territory belonging to neither of the two states. This view was supported by the continued practice of patrols of Allied soldiers of all four Allies in all four sectors. Thus, occasionally Western Allied soldiers were on patrol in East Berlin and Soviet soldiers were patrolling in West Berlin. After the Wall was built, East Germany wanted to control Western Allied patrols upon entering or leaving East Berlin, a practice that the Western Allies regarded as unacceptable. So, after protests to the Soviets, the patrols continued uncontrolled on both sides, with the tacit agreement that the western Allies would not use their patrolling privileges for helping Easterners to flee to the West.[1]

In many ways, West Berlin functioned as the *de facto* 11th state of West Germany, and was depicted on maps published in the West as being a part of West Germany. There was freedom of movement (to the extent allowed by geography) between West Berlin and West Germany. There were no separate immigration regulations for West Berlin: all immigration rules for West Germany were followed in West Berlin. West German entry visas issued to visitors were stamped with "valid for entry into the Federal Republic of Germany including Berlin (West)", authorising entry to West Berlin as well as West Germany.

West Berlin remained a military occupation zone until 3 October 1990, the day of unification of East Germany, East and West Berlin with the West German Federal Republic of Germany. The West German Federal Government, as well as the governments of most western nations, considered East Berlin to be a "separate entity" from East Germany.[2]

The ambiguous legal status of West Berlin meant that West Berliners were not eligible to vote in federal elections; instead, they were indirectly represented in the *Bundestag* by 20 non-voting delegates chosen by the West Berlin House of Representatives. Similarly, the West Berlin Senate sent non-voting delegates to the *Bundesrat*. However, as German citizens, West Berliners were able to stand for election, such as Social Democrat Chancellor Willy Brandt, who was elected via his party's list of candidates. Also, men there were exempt from the Federal Republic's compulsory military service; this exemption made the city a popular destination for West German youths, which resulted in a flourishing counterculture, which became one of the defining features of the city.

Communist countries, however, did not recognise West Berlin as part of West Germany and usually described it as a "third" German jurisdiction, called *selbständige politische Einheit* (English: *independent political entity*). On maps of East Berlin, West Berlin often did not appear as an adjacent urban area but as a monochrome terra incognita, sometimes showing the letters WB, meaning West Berlin, or overlaid with a legend or pictures.

Citizenship

While East Germany established an East German citizenship as part of its second constitution in 1967, a distinct West German citizenship did not exist. Instead, West Germany continued the definition of pre-WW2 German citizenship for all ethnic or naturalised Germans in West Germany, East Germany or any part of Berlin. So while West Berlin was not unanimously regarded as part of the Federal Republic, its citizens were treated like West German citizens by West German authorities, save for the limitations imposed by West Berlin's legal status.

This meant that West Berliners could circumvent some of these limitations if they had a second home in West Germany proper. For example, they could vote in Bundestag elections and they could be conscripted into West German military service.

Immigration

The Federal Republic of Germany issued West German passports to West Berliners on request that showed West Berlin as their place of residence. However, West Berliners could not use their passports for crossing East German borders and were denied entrance by any country of the Eastern Bloc, since governments of these countries held the view that West Germany was not authorised to issue legal papers for West Berliners. However, West Berliners travelling with West German passports carrying a secondary address in West Germany were treated as West Germans by the East German authorities.

Since West Berlin was not a sovereign state, it did not issue passports. West Berliners were issued an *auxiliary identity card* (German: *Behelfsmäßiger Personalausweis*) by the city state of Berlin (West) that was devoid of any West German federal symbols and did not indicate citizenship. From 11 June 1968, East Germany made it mandatory that West Berlin and West German *transit passengers* obtain a transit visa (German: *Transitvisum*), issued upon entering East Germany, because by its second constitution East Germany considered West Germans and West Berliners as foreigners. Since identity cards had no pages to stamp visas, the Eastern visa departments stamped their visas onto separate leaflets which were loosely stuck into the identity cards, which until the mid-1980s, were little booklets. A fee of West German Deutsche Mark (DM) 5, levied by East Germany from the *transit passengers*, could be reimbursed by the West German Federal Government.

For entering visa-requiring western countries, like the USA, West Berliners commonly used West German passports. However, for countries which did not require stamped visas for entry, including Switzerland, Austria, and many members of the then European Economic Community, West Berlin identity cards were also acceptable for entry. Occasionally, East Germany selectively banned travellers on their way through East Germany. From 13 April 1968, ministers and leading officials of the West German Federal Government were denied transit until further notice. In January 1970 East Germany interrupted transit traffic several times, because parliamentary committees of the West German Bundestag met for sessions in West Berlin, which — according to the East German authorities — they were not allowed to do, since West Berlin was not a part of West Germany.

Naming conventions

Most Westerners called the Western sectors "Berlin", unless further distinction was necessary. The West German Federal government officially called West Berlin "Berlin (West)", whereas the East German government commonly referred to it as "Westberlin"; it began to use "Berlin (West)" only in the late 1980s. Starting from 31 May 1961, East Berlin was officially called *Berlin, Capital of the GDR* (German: *Berlin, Hauptstadt der DDR*, replacing the formerly used term *Democratic Berlin*), or simply "Berlin," by East Germany, and "Berlin (Ost)" by the West German Federal government. Other names used by West German media included "Ost-Berlin", "Ostberlin", or "Ostsektor". These different naming conventions for the divided parts of Berlin when followed by individuals, governments, or media commonly indicated their political leanings.

Period following the building of the Wall

On 26 June 1963, U.S. President John F. Kennedy visited West Berlin and gave a public speech known for its famous phrase "Ich bin ein Berliner".

President John F. Kennedy addressing from Rathaus Schöneberg the people of Berlin on Rudolf-Wilde-Platz (today's John-F.-Kennedy-Platz), 26 June 1963.

The Four Power Agreement on Berlin (September 1971) and the Transit Agreement (May 1972) helped to slightly ease tensions over the status of West Berlin. While many restrictions remained in place, it also made it easier for West Berliners to travel to East Germany and it simplified the regulations for Germans travelling along the autobahn transit routes.

At the Brandenburg Gate in 1987, U.S. President Ronald Reagan provided a challenge to the then-Soviet premier: "General Secretary Gorbachev, if you seek peace, if you seek prosperity for the Soviet Union and Eastern Europe, if you seek liberalization: Come here to this gate! Mr. Gorbachev, open this gate! Mr. Gorbachev, tear down this wall!"

President Kennedy, Governing Mayor Brandt and Chancellor Adenauer (from left) standing in an open limousine in Rheinstraße on 26 June 1963, view to the city hall of Friedenau.

On 9 November 1989, the Wall was opened, and the two parts of the city were once again physically—though at this point not legally—united. The Two Plus Four Treaty, signed by the two German states and the four wartime allies, paved the way for German reunification and an end to the western occupation of West Berlin. On 3 October 1990, West Berlin and East Berlin were united as the city of Berlin, which then acceded to the Federal Republic as a *state*. West Berlin and East Berlin thus both formally ceased to exist.

Exclaves

West Berlin's border was identical with the municipal boundary of Berlin as defined in the Greater Berlin Act of 1920 and amended in 1938, and the border between the Soviet sector and the French, British and American sectors respectively, which followed the boundaries of Berlin administrative boroughs as defined in the same years. Another amendment was added in 1945 at the border between the British sector of Berlin and the Soviet zone such that the Wehrmacht airfield at Berlin-Gatow became British and the airfield at Berlin-Staaken became Soviet. The resulting borderline of all these changes was complicated with a lot of geographical oddities, including a number of exclaves and enclaves that Greater Berlin had inside some neighbouring municipalities since 1920, all of which happened to become part of the British or American sectors after 1945, thus parts of West Berlin surrounded by East Germany.

Map of divided Berlin, indicating by the broken line at Berlin's western border the territorial redeployment decided by the Allies. Five greater of West Berlin's twelve exclaves are also shown.

Furthermore the Gatow/Staaken exchange in August 1945 resulted in the geographically western half of Berlin-Staaken, which was located at the western outskirts of the city, becoming de jure Soviet occupied. However, the de facto administration remained with the Borough of Spandau in the British sector. So all inhabitants of Staaken could vote for West Berlin's city state elections in 1948 and 1950. On 1 February 1951 East German Volkspolizei surprised the West Staakeners and occupied western Staaken and ended the administration by the Spandau Borough; instead, western Staaken became an exclave of the Soviet occupied borough Berlin-Mitte in the city centre. However, on June 1, 1952, western Staaken's de facto administration was conveyed to neighbouring East German Falkensee in the East German district Nauen. This situation was undone on 3 October 1990, the day of German unification, when West Staaken was reincorporated into united Berlin.

Under the Four Power Agreement on Berlin in 1971, the Allies authorised West Berlin to negotiate territorial allocations with East Germany. On 20 December 1971, the first territorial re-allocations took place and included the exclaves numbered 1–3, 6, 8, 10 and 11 (mentioned below), connecting the latter with West Berlin and ceding the former six to East Germany as well as including a payment of four million West German Deutsche Mark to the East. The remaining exclaves were either ceded (No. 5, 7 and 12) to East Germany or territorially connected with West Berlin (No. 4 and 6) in a second redeployment in 1988.

West Berlin's twelve exclaves were the following:

- 1–3 *Böttcherberg (German)* (0.30 ha/0.74 acre): three unconnected, uninhabited and unused pieces of land, belonged to West Berlin's Borough of Zehlendorf, ceded to East Germany in 1971, since then a part of Potsdam.
- 4 *Erlengrund (German)* (0.51 ha/1.26 acre): Allotment club, seasonally inhabited, belonging to the Borough of Spandau, territorially connected with West Berlin, when East Germany ceded the adjacent tract of land in 1988. Until 1988, the members of the allotment club in West Berlin could only access it via a short passage while accompanied by East German border guards. Except for emergency rescuers, no one else was allowed to access the exclave. The path to Erlengrund was fenced on both sides so not to allow access to East Germans.
- 5 *Falkenhagener Wiese* (45.44 ha/112.28 acre): unused grassland, belonged to the Borough of Spandau, ceded to East Germany in 1988, since then a part of Falkensee.
- 6 *Fichtewiese (German)* (3.51 ha/8.67 acre): Allotment club, seasonally inhabited, belonging to the Borough of Spandau, and territorially connected with West Berlin, when East Germany ceded the adjacent tract of land in 1988. Until 1988, the allotment holders had to pass East German border controls on their way between Fichtewiese and the rest of West Berlin. Except for emergency rescuers, no one else was allowed to access the exclave. The path connecting Fichtewiese with West Berlin was fenced on both sides so not to allow access to East Germans.
- 7 *Finkenkrug*, (3.45 ha/8.53 acre): inhabited by East Germans, five km away from West Berlin's border, and belonged to the Borough of Spandau, ceded to East Germany in 1971, since then a part of Falkensee.
- 8 *Große Kuhlake* (8.03 ha/19.84 acre): unused grassland, belonged to the Borough of Spandau, ceded to East Germany in 1971.
- 9 *Laßzins-Wiesen* (13.49 ha/33.33 acre): unused grassland, belonged to the Borough of Spandau, ceded to East Germany in 1988, since then a part of Schönwalde.
- 10 *Nuthewiesen (German)* (3.64 ha/8.99 acre): uninhabited wet meadows, belonged to the Borough of Zehlendorf, ceded to East Germany in 1971, since then a part of Potsdam.
- 11 *Steinstücken* (12.67 ha/31.31 acre): inhabited by West Berliners and belonging to the Borough of Zehlendorf. It was territorially connected with West Berlin, when East Germany ceded a small strip of land in 1971. Until 1971, the inhabitants had to pass East German border controls on their way between Steinstücken and the main area of West Berlin. Except for emergency rescuers and repair personnel, no one else was allowed to access the exclave. The road connecting Steinstücken was immured on both sides not allowing Easterners to enter it.
- 12 *Wüste Mark (German)* (21.83 ha/53.94 acre): despite its name (meaning "desert border region" in English), not a wasteland, but a seasonally tilled acreage, belonging to the Borough of Zehlendorf, ceded to East Germany in 1988, and since then a part of Stahnsdorf. Wüste Mark is a tract of land adjacent to Wilmersdorf's forest cemetery

in Güterfelde. Until 1988, the West Berlin farmer tilling the land was allowed after filing a formal request to cross with his tractor through East Germany.

Transport and transit travel

West Berliners could travel to West Germany and all Western and non-aligned states at all times, except of the period of the Berlin Blockade by the Soviet Union (24 June 1948 to 12 May 1949), due to restrictions on passenger flight capacity imposed by the airlift. Travelling to and from West Berlin by road or train always required passing through East German border checks, since West Berlin was an enclave surrounded by East Germany and East Berlin.

Road traffic

For travel from West Berlin through East Germany by car or rail a valid passport was required for citizens of West Germany and other western nationals to be produced at East German border checks; West Berliners could only get admission through their identity cards (see above). For travel from West Berlin to Denmark, West Germany, or Sweden, via dedicated East German *transit routes* (German: *Transitstrecke*), East German border guards issued a *transit visa* for a fee of 5 Western Deutsche Mark. For journeys between West Berlin and Poland or Czechoslovakia through East Germany, each traveller was also required to present a valid visa for the destination country.

The *transit routes* for road travel connecting West Berlin to other destinations usually consisted of autobahns and other highways, marked by *Transit* signs. *Transit travellers* (German: *Transitreisende*) were prohibited to leave the *transit routes*, and occasional traffic checkpoints would check for violators.

There were four *transit routes* between West Berlin and West Germany:

- One between West Berlin's *Heerstraße* with the East German checkpoint in Dallgow until 1951, then replaced by Staaken for destinations in Northern Germany (originally via highway *F 5*) at the Eastern checkpoint in Horst (a part of today's Nostorf) and the Western Lauenburg upon Elbe. These were replaced in 20 November 1982 by a new autobahn crossing at Zarrentin (E)/Gudow (W).[3] On 1 January 1988, the new Stolpe checkpoint opened on this route to West Berlin. This is part of today's Hohen Neuendorf (E)/Berlin-Heiligensee (W).
- A second transit route led to Northwestern and Western Germany — following today's A 2 — crossing the inner German border at Marienborn (E)/Helmstedt (W), also called *Checkpoint Alpha*.
- A third route to Southwestern Germany consisted of today's A 9 and A 4 with border crossing at Wartha (E)/Herleshausen (W),.
- A fourth (via today's A 9) to Southern Germany had border crossings originally at Mount Juchhöh (E)/Töpen (W) and later at Hirschberg upon Saale (E)/ Rudolphstein (a part of today's Berg in Upper Franconia) (W).

The latter three routes used autobahns built during the Nazi era. They left West Berlin at *Checkpoint Dreilinden*, also called Checkpoint Bravo (W)/Potsdam-Drewitz (E). *Transit routes* to Poland were via today's A 11 to Nadrensee-Pomellen (East Germany, GDR)/Kołbaskowo (Kolbitzow) (PL), eastwards via today's A 12 to Frankfurt upon Oder (GDR)/Słubice (PL), or southeastwards via today's A 13 and A 15 to Forst in Lusatia/Baršć (GDR)/Zasieki (Berge) (PL). Additional routes led to Denmark and Sweden by ferry between Rostock (GDR) and Gedser (DK) and by ferry between Sassnitz (GDR) and Rønne (DK) or Trelleborg (S). Routes to Czechoslovakia were via Schmilka (GDR)/Hřensko (Herrnskretschen) (ČSSR) and via Fürstenau (a part of today's Geising) (GDR)/Cínovec (Cinvald/Böhmisch Zinnwald) (ČSSR).

East German border crossing Potsdam-Drewitz on 31 March 1972: Applying eastern lead seals to western trucks, entering the transit route, in order to prevent potential Eastern German escapees from hiding in the cargo room.

The *transit routes* were also used for East German domestic traffic. This meant that transit passengers could potentially meet with East Germans and East Berliners at restaurants at motorway rest stops. Since such meetings were deemed illegal by the East German government, border guards would calculate the travel duration from the time of entry and exit of the transit route. Excessive time spent for transit travel could arouse their suspicion and prompt questioning or additional checking by the border guards. Western coaches could stop only at dedicated service areas, since the East German government was concerned that East Germans might potentially use coaches to escape into the West.

On 1 September 1951 East Germany, because of shortages in foreign currencies, started to levy road tolls from cars using the *transit routes*. At first the toll amounted to Eastern Deutsche Mark 10 per passenger car and 10 to 50 for trucks, depending on size. Eastern Deutsche Marks had to be exchanged for Western Deutsche Mark a rate of 1:1. On 30 March 1955, East Germany raised the toll for passenger cars to 30 Deutsche Marks, but after West German protests, in June of the same year it changed it back to the previous rate. Following a new agreement between East and West Germany, starting from 1 January 1980 the Western *Federal Government* paid an annual lump sum (German: *Transitpauschale*) of 50 million Western Deutsche Marks to the Eastern government, so that *transit passengers* no longer had to pay tolls individually.

Railway

Four transit train connections—earlier also called *interzonal train* (German: *Interzonenzug*)—connected West Berlin with Hamburg via Schwanheide (E)/Büchen (W) in the North, with Hanover via Marienborn (E)/Helmstedt (W) in the West, with Frankfurt upon Main via Gerstungen (E)/Hönebach (W) in the Southwest, and with Nuremberg via Probstzella (E)/Ludwigsstadt (W) in the South of West Germany. These transit trains did not service domestic passengers of East Germany and made stops in East Germany almost only for East German border guards upon entering and leaving of the country. Until the construction of the Berlin Wall, *interzonal trains* would also stop once on their way within East Germany for travellers having a visa for entering or leaving East Germany. Train travel from West Berlin to Czechoslovakia, Denmark (by ferry), Poland, or Sweden (by ferry) required a visa to enter East Berlin or East Germany to allow transfer to an international train—which also carried domestic passengers—bound for an international destination. One railway connection between West Berlin and Oebisfelde (E)/Wolfsburg (W) was reserved for freight trains only.

In July and August 1945, the three Western Allies and the Soviet Union decided that the railways, previously serviced by the Deutsche Reichsbahn (German National Railway), should continue to be operated by one railway administration to service all four sectors. So West Berlin had — with the exception of a few small private railway lines — no separate railway administration. Furthermore, the operation of the Reichsbahn's Berlin S-Bahn electric metropolitan transport network, consisting of commuter trains, was also maintained. After the founding of East Germany on 7 October 1949 it gained responsibility for the Reichsbahn in its territory. East Germany continued to run its railways under the official name Deutsche Reichsbahn, which thus maintained responsibility for almost all railway transport in all four sectors of Berlin.[4] The GDR-controlled 'Bahnpolizei', the Reichsbahn's railway police, were authorised to patrol station premises and other railway property in the whole city including West Berlin.

After the Berlin Blockade *transit trains* (German: *Transitzüge*) would leave and enter West Berlin only via one line through Berlin-Wannsee railway station (W) and Potsdam Griebnitzsee railway station (E). All *transit trains* would start or end in East Berlin, passing through West Berlin with only one stop in the Western Berlin Zoologischer Garten railway station, which became West Berlin's main railway station. Until 1952, the Reichsbahn also permitted stops at other stations on the way through the Western sectors. After easing of tensions between East and West Germany, starting on 30 May 1976 *transit trains* going westwards, southwestwards, or southwards stopped once again at station of Wannsee. For *transit trains* going northwestwards, a shorter line was reopened on 26 September 1976 with an additional stop at the then Berlin-Spandau railway station, entering East Germany at Staaken. Many Reichsbahn employees working in West Berlin were West Berliners. Their East German employer, whose proceeds from ticket sales for Western Deutsche Marks contributed to East Germany's foreign revenues, tried to hold down wage social security contributions in Western Deutsche Mark. Therefore, West Berlin employees of the Reichsbahn were paid partly in Eastern German currency. However, they could spend this money in East Germany and take their purchases to West Berlin, which other Westerners could not do to the same extent. West Berlin employees were trained in East Germany and employed under East German labour laws.[5] West Berliners employed by the Reichsbahn were not included in the Western health insurance system. The Reichsbahn ran its own hospital for them in West Berlin, the building of which is now used as the headquarters of Bombardier Transportation. For certain patients, the Reichsbahn would facilitate treatment in a hospital in East Berlin. In medical emergencies, the employees could use West Berlin doctors and hospitals, which would then be paid for by the Reichsbahn.

The GDR used the western stations to distribute propaganda and display posters with slogans like "Americans Go Home." On 1 May, May Day, a state holiday in East and West, S-Bahn trains were sometimes decorated with the East German banner and a red flag.

Waterways

Two waterways via the rivers and canals Havel and Mittellandkanal were open for inland navigation, but only freight vessels were allowed to cross from West Berlin into East German waters. The Havel crossed at the East German border in Nedlitz (a part of Potsdam-Bornstedt), continuing through the Elbe-Havel Canal and then either taking the Elbe northwestwards crossing the border again at Cumlosen (E)/Schnackenburg (W) or westwards following the Mittellandkanal to Buchhorst (Oebisfelde) (E)/Rühen (W). Western freight vessels could stop only at dedicated service areas, because the East German government wanted to prevent any East Germans from boarding them. Through these waterways, West Berlin was connected with the western European inland navigation network, connecting to seaports like Hamburg and Rotterdam, as well as industrial areas such as the Ruhr Area, Mannheim, Basel, Belgium and eastern France.

In July and August 1945, the Western Allies and the Soviet Union decided that the operation and maintenance of the waterways and locks, which were previously run by the national German directorate for inland navigation (German: *Wasser- und Schifffahrtsamt Berlin*), should be continued and reconstructed in all four sectors. So, except of some later built canals and locks, West Berlin had no separate inland navigation authority, but the East Berlin-based authority operated and maintained most waterways and locks.

The western entrance to the Teltowkanal, connecting several industrial areas of West Berlin for heavy freight transport, was blocked by East Germany in Potsdam-Klein Glienicke. Therefore, vessels going to the Teltowkanal had to take a detour via the river Spree through West and East Berlin's city centre to enter the canal from the East. On 20 November 1981, East Germany reopened the western entrance, which required two more vessel border checkpoints — Dreilinden and Kleinmachnow — because the waterway crossed the border between East Germany and West Berlin four times. Another transit waterway connected West Berlin via the East German vessel checkpoint at Hennigsdorf and the Oder-Havel Canal with the Oder river and Polish Szczecin (Stettin).

Air traffic

Eastern refugees boarding a plane at Tempelhof Airport to fly into West Germany, 1953.

Air traffic was the only connection between West Berlin and the Western world that was not directly under East German control. British European Airways opened the first regular service for civilians on 4 July 1948 between West Berlin and Hamburg. Tickets were originally sold for pounds sterling only. West Berliners and West Germans who had earlier fled East Germany or East Berlin, and thus could face imprisonment on entering East Germany or East Berlin, could only take flights for travel to and from West Berlin.[6] To enable individuals threatened by East German imprisonment to fly to and from West Berlin the West German government subsidized the flights.

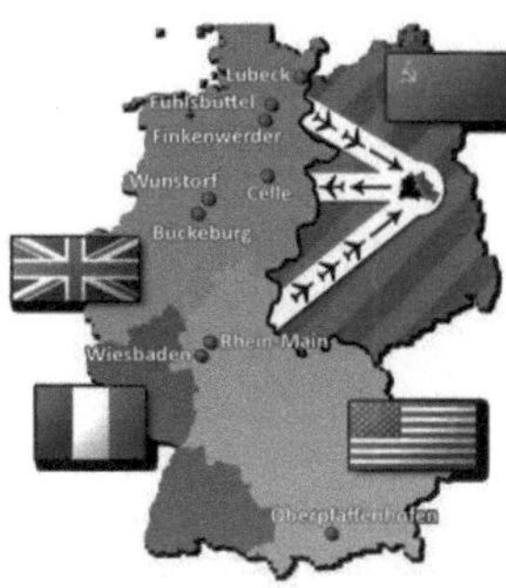

The only three permissible West Berlin Air Corridors.

The flights between West Germany and West Berlin were under Allied control by the quadripartite Berlin Air Safety Center. According to permanent agreements, three air corridors to West Germany were provided, which were open only for British, French or U.S. military planes or civilian planes registered with companies in those countries.

The airspace controlled by the *Berlin Air Safety Center* comprised a radius of 20 miles (32.12 km) around the seat of the Center in the Kammergericht building in Berlin-Schöneberg — thus covering most of East and West Berlin and the three corridors, of the same width — one northwestwards to Hamburg-Fuhlsbüttel Airport, one westwards to Hanover, one southwestwards to Frankfurt upon Main (Rhein-Main Air Base). Also the airspace expanding to a width of 20 miles (32 km) over the German-German border was subject to the control by the *Berlin Air Safety Center.*

The West German Lufthansa and most other airlines were not permitted to fly to West Berlin. Flights of Lufthansa or the East German Interflug servicing connections between East and West Germany (such as between West German Düsseldorf and Hamburg and East German Leipzig) began in August 1989, but these routes had to go through Czechoslovak or Danish airspace.

Traffic between West Berlin and East Germany

Until 1953, travelling from West Berlin into East Germany (German Democratic Republic (GDR)), fell under Interzonal traffic regulations overseen by three Allied military governments (the Soviet Military Administration in Germany (SVAG), the Control Commission for Germany – British Element, and the Office of Military Government/United States (OMGUS)). On 27 May 1952, East Germany closed its border with West Germany and its 115-kilometre (71 mi)-long border with West Berlin. From then on West Berliners required a permit to enter East Germany. East German border checkpoints were established in East German suburbs of West Berlin, and most streets were gradually closed for interzonal travel into East Germany. The last checkpoint to remain open was located at the Glienicker Brücke towards Potsdam, until it was also closed by East Germany on 3 July 1953. The checkpoint at Staaken's Heerstraße remained open only for transit traffic to West Germany.

This caused hardship for many West Berlin residents, especially those who had friends and family in East Germany. However, East Germans could still enter West Berlin. A number of cemeteries located in East Germany were also affected by the closure. Many church congregations in Berlin owned cemeteries outside the city, so many West Berlin congregations had cemeteries that were located in East Germany. For example, the *Friedhof vor*

Charlottenburg (in English: *Cemetery in front/outside of Charlottenburg*) was located in the East German suburb of Dallgow, yet belonged to Catholic congregations in Berlin-Charlottenburg. So many West Berliners wishing to visit the grave of a relative or friend on cemeteries located in East Germany were now unable to do so. Until 1961, East Germany sparsely issued permits to West Berliners to visit the cemeteries on the Catholic feast of *All Saints* on 1 November and on the Protestant Day of Repentance and Prayer.

In 1948–1952, the Reichsbahn connected the western suburbs of West Berlin to its S-Bahn network. Train routes servicing these suburbs formerly went through West Berlin stations, but ceased to make stops in the western stations or terminated service before entering West Berlin. Private West Berlin railway lines like the *Neukölln-Mittenwalder Eisenbahn* (NME), connecting the East German Mittenwalde with West Berlin-Neukölln and the *Bötzowbahn* between West Berlin-Spandau and East German Hennigsdorf, were disrupted at the border between West Berlin and East Germany on 26 October 1948 and August 1950, respectively. Tramways and bus routes that connected West Berlin with its East German suburbs and were operated by West Berlin's public transport operator Berliner Verkehrsbetriebe Gesellschaft (*BVG*-West) ceased operation on 14 October 1950, after West Berlin tram and bus drivers had been repeatedly stopped and arrested by East German police. The BVG (West) terminated route sections that extended into East Germany, like the southern end of tram line 47 to Schönefeld, the southwestern end of tram line 96 to Kleinmachnow, as well as two bus lines to Glienicke at the Nordbahn, north, and to Falkensee, northwest of West Berlin.

The Reichsbahn shut down all of its West Berlin terminal stations and redirected its trains to stations in East Berlin, starting with Berlin Görlitzer Bahnhof — closed on 29 April 1951 — before serving rail traffic with Görlitz and the southeast of East Germany. On 28 August 1951, trains usually serving Berlin Lehrter Bahnhof were redirected to stations in East Berlin, while trains from West Germany were redirected to the Western *Berlin Zoologischer Garten*. The Reichsbahn also closed down both Berlin Anhalter Bahnhof and Berlin Nordbahnhof, on 18 May 1952.

On 28 August 1951, the Reichsbahn opened a new connection — from Spandau via Berlin Jungfernheide station — for the S-Bahn lines connecting East German suburbs to the west of West Berlin (namely Falkensee, Staaken) with East Berlin, thus circumventing the centre of West Berlin. In June 1953, the Reichsbahn further cut off West Berlin from its East German suburbs by the introduction of additional *express S-Bahn trains* (German: *Durchläufer*). These routes originated from several East German suburbs bordering West Berlin (such as Falkensee, Potsdam, Oranienburg, Staaken, and Velten), crossing West Berlin non-stop until reaching its destinations in East Berlin. However, also the regular S-Bahn connections with West Berlin's East German suburbs, stopping at every Western station, continued. From 17 June to 9 July 1953, East Germany blocked off any traffic between East and West due to the Uprising of 1953 in East Germany.

From 4 October 1953, all S-Bahn trains crossing the border between East Germany and Berlin had to pass a border checkpoint in East Germany. Travellers from East Germany were checked before entering any part of Berlin, to identify individuals intending to escape into West Berlin or smuggling rationed or rare goods into West Berlin. S-Bahn trains were checked at Hoppegarten, Mahlow and Zepernick in East Germany bordering East Berlin and in Hohen Neuendorf, Potsdam-Griebnitzsee and Staaken-Albrechtshof in East Germany bordering West Berlin. On 4 June 1954, the *Bahnhof Hennigsdorf Süd* station located next to West Berlin was opened solely for border controls, also to monitor West Berliners entering or leaving East Berlin, which they could still do freely, while they were not allowed to cross into East Germany proper without a special permit.

In 1951, the Reichsbahn began construction work on the *Berlin outer-circle railway line* (German: *Berliner Außenring (BAR)*). This circular line connected all train routes heading for West Berlin and accommodated all domestic GDR traffic, thus directing railway traffic into East Berlin while by-passing West Berlin. Commuters in the East German suburbs around West Berlin now boarded *Sputnik* express trains, which took them into East Berlin without crossing any western sectors. With the completion of the Außenring, there was no further need for express S-Bahn trains crossing the West Berlin border and thus their service ended on 4 May 1958, while stopping S-Bahn trains continued service. However, while East Germans could get off in West Berlin, West Berliners needed the hard-to get permits to enter East Germany by S-Bahn. With the construction of the Berlin Wall on 13 August 1961, any remaining railway traffic between West Berlin and its East German suburbs ended. The rail traffic between East and West Berlin was sharply reduced and restricted to few checkpoints under GDR control. East Berliners and East Germans were then unable to freely enter and leave West Berlin. However, international visitors could obtain visas for East Berlin upon crossing one of the checkpoints at the Wall.

Following the policy of détente of the *Federal Government* under Chancellor Willy Brandt, West Berliners could again apply for visas to visit East Germany, which were granted more freely than in the period until 1961. On 4 June 1972, West Berlin's public transport operator *BVG* could open its first bus line into the East German suburbs since 1950 (line E to Potsdam via *Checkpoint Bravo* as it was known to the US military). This route was open only to persons bearing all the necessary East German permits and visas. For visits to East Germany, West Berliners could use four checkpoints along the East German border around West Berlin: The two road transit checkpoints Dreilinden (W)/Drewitz (E) and Berlin-Heiligensee (W)/Stolpe (E) as well as the old transit checkpoint at Heerstraße (W)/Staaken (E) and the checkpoint at Waltersdorfer Chaussee (W)/Schönefeld (E), which was also open for travellers boarding international flights at Schönefeld Airport.

Traffic between East and West Berlin

While East and West Berlin became formally separate jurisdictions in September 1948, and while there were travel restrictions in all other directions, for more than a decade, freedom of movement existed between the western sectors and the eastern sector of the city. However, time and again Soviet and later East German authorities imposed temporary restrictions for certain persons, certain routes, and certain means of transport. Gradually the eastern authorities disconnected and separated the two parts of the city.

While the Soviets blocked all transport to West Berlin (Berlin Blockade between 24 June 1948 to 12 May 1949) they increased the supplies for food in East Berlin in order to gain the compliance of West Berliners who at that time still had free access to East Berlin. West Berliners buying food in East Berlin were regarded as approving of the Soviet attempt to repress the Western Allies from West Berlin. This was seen as support by the communists and as treason by most Westerners. Until that time all over Germany food and other necessary supplies had been available only with ration stamps issued by one's municipality, this was until the Communist putsch in Berlin's city government in September 1948 – the unitary City Council of Greater Berlin (German: *Magistrat von Groß Berlin*) for East and West.

By July 1948 a mere 19,000 West Berliners out of a total of almost 2 million covered their food requirements in East Berlin. So 99% of the West Berliners preferred to live with shorter supplies than before the Blockade but support the Western Allies' position. In West Germany rationing of most products had ended with the introduction of the Western Deutsche Mark on 21 June 1948. The new currency was also introduced in West Berlin on 24 June and this, at least officially was the justification for the Soviet Blockade due to which, rationing in West Berlin had to continue. However in the course of the Berlin Air Lift some supplies were increased beyond the pre-Blockade level and therefore certain rations in West Berlin were raised.

While West Berliners were officially welcome to buy food in East Berlin, the Soviets tried to prevent them buying other essential supplies there, particularly coal and fuel. For this reason, on 9 November 1948, they opened checkpoints on 70 streets entering West Berlin and closed the others for horse carriages, lorries and cars, later (16 March 1949) the Soviets erected roadblocks on the closed streets. From 15 November 1948 West Berlin ration stamps were no longer accepted in East Berlin. All the same, the Soviets started a campaign with the slogan *The smart West Berliner buys at the HO* (German: *Der kluge West-Berliner kauft in der HO*), the HO being the Soviet zone chain of shops. They also opened so-called "Free Shops" in the Eastern Sector, offering supplies without ration stamps, but at extremely high prices in Eastern Deutsche Marks. Ordinary East and West Berliners could only afford to buy there if they had revenues in Western Deutsche Mark and bartered the needed Eastern Deutsche Mark on the spontaneous currency markets, which developed in the British sector at the Zoo station. Their demand and supply determined a barter ratio in favour of the Western Deutsche Mark with more than 2 Eastern Deutsche Marks offered for one Western Deutsche Mark. After the Blockade – when holders of Western Deutsche Marks could buy as much they could afford, up to five and six east marks were offered for one west mark. In the East however, the Soviets had arbitrarily decreed a rate of 1 for 1 and exchanging at other rates was criminalised.

On 12 May 1949 the Blockade ended and all roadblocks and checkpoints between East and West Berlin were removed. The Berlin Airlift, however, continued until 30 September 1949 to amass sufficient supplies in West Berlin in readiness for another possible blockade, ensuring that an airlift could then be re-started with ease. On 2 May 1949 the power stations in East Berlin again started to supply West Berlin with sufficient electricity, which had to be rationed to some hours a day after the usual supplies had been interrupted at the start of the Blockade. However, the Western Allies and the West Berlin City Council decided to be self sufficient in terms of electricity generation capacity, to be independent of Eastern supplies and not to be held to ransom by the eastern authorities. On 1 December 1949 the new *powerhouse West* (German: *Kraftwerk West*, in 1953 renamed after the former Governing Mayor of West Berlin into *Kraftwerk Reuter West*) went on line and West Berlin's electricity board declared independence from Eastern supplies. However, for a time Eastern electricity continued to be supplied albeit intermittently. Supply was interrupted from 1 July until the end of 1950 and then started again until 4 March 1952, when the East finally switched it off. From then on West Berlin turned into an 'electricity island' within a pan-European electricity grid that had developed from the 1920s, because electricity transfers between East and West Germany never fully ceased. The 'electricity island' situation was noticed most in situations of particularly high demand; in other areas of Europe peaks in demand could be met by tapping into electricity supplies from neighbouring areas, but in West Berlin this was not an option and for certain users the lights would go out.

In 1952 West Berliners were restricted entry to East Germany proper by means of a hard-to-obtain East German permit. Free entry to East Berlin remained possible until 1961 and the building of the Wall. Berlin's underground (Untergrundbahn, U-Bahn) and Berlin's S-Bahn (a metropolitan public transit network), rebuilt after the war, continued to span all occupation sectors. Many people lived in one half of the city and had family, friends, and jobs in the other. However, the East continuously reduced the means of public transport between East and West, with private cars being a very rare privilege in the East and still a luxury in the West.

Starting on 15 January 1953 the tram network was interrupted. East Berlin's public transport operator *Berliner Verkehrsbetriebe* (*BVG*-East, *BVB* as of 1 January 1969) staffed all trams, whose lines crossed the sectorial border, with women drivers, which were not permitted as drivers by the *BVG* (West), West Berlin's public transport operator. Instead of changing the Western rules, so that the Easterly intended interruption of the cross-border tram traffic would not happen, the *BVG* (West) insisted on male drivers. So cross-border tram traffic ended on 16 January.[7] In East German propaganda this was a point for the East, arguing that the West did not allow drivers coming with their trams from the East to continue along their line into the West, but remaining silent on the fact that the end of cross-border tram traffic was most welcome to the East. The underground and the S-Bahn networks, except the above-mentioned *traverse S-Bahn trains*, continued to provide services between East and West Berlin. However, occasionally the East Berlin police – in the streets and on cross-border trains in East Berlin – identified suspicious behaviour (such as carrying heavy loads westwards) and watched out for unwelcome Westerners.

Once in a while West Germans were banned from entering East Berlin. This was the case between 29 August and 1 September 1960, when so-called *homecomers* (German: *Heimkehrer*) from all around West Germany and West Berlin met for a convention in that city. The *homecomers* from a long compulsory detention in the Soviet Union, with their special experiences there, were highly unwelcome in East Berlin.[8] Since they could not be recognised by their identification papers all West Germans were banned for these days from East Berlin. West Berliners were allowed, since the quadripartite Allied status quo provided for their free movement around all four sectors. From 8 September 1960 on, the East subjected all West Germans to apply for a permit before entering East Berlin.

As the communist government in the East gained tighter control, and the economic recovery in the West significantly outperformed the Eastern development, more than a hundred thousand East Germans and East Berliners left East Germany and East Berlin for the West every year. East Germany closed the borders between East and West Germany and sealed off the border with West Berlin in 1952; but because of the quadripartite Allied status of the city, the 46-kilometre (29 mi)-long sectorial border between East and West Berlin remained open. As there was freedom of movement between West Berlin and West Germany, Easterners could use the city as a transit point to West Germany, usually travelling there by air.

To stop this drain of people defecting, the East German government built the Berlin Wall, thus physically closing off West Berlin from East Berlin and East Germany, on 13 August 1961. All Eastern streets, bridges, paths, windows, doors, gates or sewers opening to West Berlin were systematically sealed off by walls, concrete elements, barbed wire or bars. The Wall was directed against the Easterners, who by its construction were no longer allowed to leave the East, except with an Eastern permit, not usually granted.

Westerners were still granted visas on entering East Berlin. Initially eight street checkpoints were opened, and one checkpoint in the Berlin Friedrichstraße railway station, which was reached by one line of the Western underground (today's U 6), two Western S-Bahn lines, one under and one above ground (approximate to today's S 2 and S 3, however lines changed a lot from 1990 onwards), and *transit trains* between West Germany and West Berlin started and ended there.

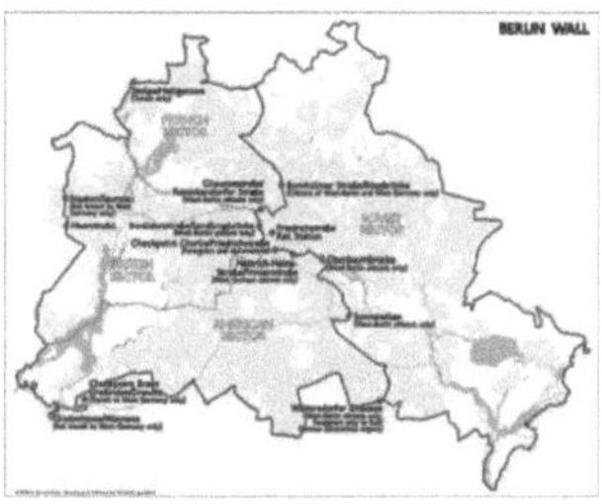

Map showing location of the Berlin Wall and transit points

The eight street checkpoints were – from North to South along the Wall – on Bornholmer Straße, Chausseestraße, Invalidenstraße, Brandenburg Gate, Friedrichstraße (Checkpoint Charlie in US military denomination, since this crossing was to their sector), Heinrich-Heine-Straße (also Checkpoint Delta), and Sonnenallee.

When the construction of the Wall started after midnight early on 13 August, West Berlin's Governing Mayor Willy Brandt was on a West German federal election campaigning tour in West Germany. Arriving by train in Hanover at 4a.m. he was informed about the Wall and flew back to West Berlin's Tempelhof Central Airport.

An eastern water cannon vehicle directed at western protesters in front of the Brandenburg Gate, August 1961

In the course of the day he protested along with many other West Berliners on Potsdamer Platz and at the *Brandenburg Gate*. On 14 August, under the pretext that Western demonstrations required it, the East closed the checkpoint at the *Brandenburg Gate*, 'until further notice', a situation that was to last until 22 December 1989, when it was finally reopened.

On 26 August 1961 East Germany generally banned West Berliners from entering the Eastern sector. West Germans and other nationals however could still get visas on entering East Berlin. Since intra-city phone lines had been cut by the East already in May 1952 (see below) the only remaining way of communication with family or friends on the other side was by mail or at meeting in a motorway restaurant on a *transit route*, because the *transit traffic* remained unaffected throughout.

On 18 May 1962 East Germany opened the so-called Tränenpalast checkpoint hall (English: *Palace of Tears*) at Berlin Friedrichstraße station, where Easterners had to say a sometimes tearful farewell to returning Westerners as well as the few Easterners who had managed to get a permit to visit the West. Until June 1963 the East deepened its border zone around West Berlin in East Germany and East Berlin by clearing existing buildings and vegetation to create an open field of view, sealed off by the Berlin Wall towards the West and a second wall or fence of similar characteristics to the East, observed by armed men in towers, ordered to shoot possible refugees.

Finally in 1963 West Berliners were again allowed to visit East Berlin. On this occasion a further checkpoint for pedestrians only was opened on the Oberbaumbrücke. West Berliners were granted visas for a one-day-visit between 17 December 1963 and 5 January the following year. 1.2 million out of a total 1.9 million West Berliners visited East Berlin during this period. In 1964, 1965 and 1966 East Berlin was opened again to West Berliners, but each time only for a limited period.

Western police awaiting an eastern border controller at the opening of a new pedestrian border crossing. View into the vaults of Oberbaumbrücke, 21 December 1963.

West Berliners entering East Berlin at the border crossing Chausseestraße on 28 December 1963 after having been banned from visiting the eastern sector for more than two years.

East Germany found particular joy in playing with the different legal statuses it assigned to East Germans, East Berliners, West Germans, and West Berliners, as well as citizens from other countries in the world. Until 1990 East Germany designated each Border crossings in East Berlin for certain categories of persons, with only one street checkpoint being open simultaneously for West Berliners and West Germans (Bornholmer Straße) and *Berlin Friedrichstraße railway station* being open for all travellers.

On 9 September 1964, the East German Council of Ministers (government) decided to allow Eastern pensioners to visit family in West Germany or West Berlin. According to the specified regulations valid from 2 November on Eastern pensioners could apply, and were usually allowed, to travel into the West to visit relatives once a year for a maximum of four weeks. If pensioners decided not to return, the government did not miss them as manpower, unlike younger Easterners, who were subject to a system of labour and employment, which demanded almost everybody work in the Eastern command production system.

On 2 December 1964 East Germany, always short of hard currency, decreed that every Western visitor had to buy a minimum of 5 Eastern *Mark der Deutschen Notenbank* per day (MDN,[9] 1964–1968 the official name of the East German mark, to distinguish it from the West Deutsche Mark) at the still held arbitrary compulsory rate of 1:1. The five marks had to be spent, as exporting Eastern currency was illegal, which is why importing it after having bargained for it at the currency market at Zoo station was also illegal. Western pensioners and children were spared from the compulsory exchange (officially in German: *Mindestumtausch*, i.e. *minimum exchange*). Not long after East Germany held the first cash harvest from the new compulsory exchange rules by allowing West Berliners to visit East Berlin once more for a day during the Christmas season. The following year, 1965, East Germany opened the travelling season for West Berliners on 18 December. In 1966 it opened for a second harvest of Western money between the Easter (10 April) and Pentecost (29 May) holidays and later again at Christmas.

The situation only changed fundamentally after 11 December 1971 when, representing the two German states, the Western Egon Bahr and the Eastern Michael Kohl signed the Transit Agreement. This followed by a comparable agreement for West Berliners, once more allowing regular visits to East Germany and East Berlin.

After ratification of the Agreement and specifying the pertaining regulations West Berliners could apply for the first time again for visas for any chosen date to East Berlin or East Germany from 3 October 1972 onwards. If granted, a one-day-visa entitled them to leave the East until 2a.m. the following day. West Berliners were now spared the visa fee of 5 Western Deutsche Marks, not to be confused with the compulsory exchange amounting to the same sum, but yielding in return 5 Eastern marks. This financial relief did not last long, because on 15 November 1973 East Germany doubled the compulsory exchange to 10 Eastern marks, payable in West German Deutsche Marks at par.

One-day-visas for East Berlin were now issued in a fast procedure on entering East Berlin; visas for longer stays and visas for East Germany proper needed a prior application, which could be a lengthy procedure. To ease the application for West Berliners seeking such Eastern visas, the GDR Foreign Ministry was later allowed to open *Offices for the Affairs of Visits and Travelling* (German: *Büros für Besuchs- und Reiseangelegenheiten*) in West Berlin, but were not allowed to show any official symbols of East Germany. The Eastern officials working commuted every morning and evening between East and West Berlin. Their uniforms showed no official symbols except the name *Büro für Besuchs- und Reiseangelegenheiten.* They accepted visa applications and handed out confirmed visas issued in the East, to the West Berlin applicants. A shed formerly housing one such *Büro für Besuchs- und Reiseangelegenheiten* can be found on Waterlooufer 5–7 in Berlin-Kreuzberg, close to Hallesches Tor underground station.[10] The disagreement about Berlin's status was one of the most important debates of the Cold War.

Another form of traffic between East and West Berlin was the transfer of West Berlin's sewage into East Berlin and East Germany through the sewer pipes built in the late 19th and early 20th century. The sewage flowed into the East because most of the pre-war premises for sewage treatment, mostly sewage farms, happened to be in the East after the division of the city. Sewer pipes, however, once discovered as a way to flee the East, were blocked by bars. West Berlin paid for the treatment of its sewage in Western Deutsche Marks which were desperately needed by the Eastern government. Since the methods used in the East did not meet Western standards, West Berlin increased the capacity of modern sewage treatment within its own territory, so that the amount of its sewage treated in the East had considerably reduced by the time the Wall came down.

Similar was the situation with refuse. The removal, burning or disposal of the ever-growing amount of West Berlin's rubbish became a costly problem, but here too an agreement was found, since West Berlin would pay in Western Deutsche Marks. On 11 December 1974 East Germany and West Berlin's garbage utility company *BSR* signed a contract to dispose of refuse on a dump right beside the Wall in East German Groß-Ziethen (today a part of Schönefeld). An extra checkpoint, solely open for Western bin lorries was opened there. Later a second dump, further away, was opened in Vorketzin, a part of Ketzin.

As for the S-Bahn, operated in all of Berlin by the East German Reichsbahn, the construction of the Wall meant a deep cut into its integrated network of lines, especially for Berlin's circular S-Bahn line around all of the Western and Eastern inner city. The lines were separated and those mostly located in West Berlin were continued, but only accessible from West Berlin with all access in East Berlin closed. However, even before the Wall had been built, West Berliners increasingly refrained from using the S-Bahn, since boycotts against it were issued, the argument being that every S-Bahn ticket bought provided the GDR government with valuable Western Deutsche Marks.

Usage dropped further as the Western public transport operator *BVG* (West) offered parallel bus lines and expanded its network of underground lines. After the construction of the Wall usage dropped so much that running the S-Bahn lines in West Berlin turned into a loss-making exercise: wages and maintenance costs – however badly it was carried out outdid the proceeds from ticket sales. So the Reichsbahn finally agreed to surrender operation of the S-Bahn in West Berlin, as had been determined by all Allies in 1945, and on 29 December 1983 the Allies, the Senate of Berlin (West; i.e. the city state government) and the Reichsbahn signed an agreement to change the operator from Reichsbahn to *BVG* (West) which took effect on 9 January 1984.

East Berliners, just having passed the now open eastern checkpoint Bornholmer Straße, passing Bösebrücke into the French sector of Berlin on 18 November 1989.

On 9 November 1989 East Germany opened the borders for East Germans and East Berliners, who could then freely enter West Berlin. West Berlin itself had never restricted their entry. For West Berliners and West Germans the opening of the border for free entry lasted longer. The regulation concerning one-day-visas on entering the East and the compulsory minimum exchange of 25 Western Deutsche Marks by 1989, continued. However, more checkpoints were opened. Finally on 22 December 1989 East Germany granted West Berliners and West Germans free entry without charge at the existing checkpoints, demanding only valid papers. Eastern controls were slowly eased into spot checks and finally abolished on 30 June 1990, the day East and West introduced the *union concerning currency, economy and social security* (German: *Währungs-, Wirtschafts- und Sozialunion*).

Traffic between different parts of West Berlin crossing the East

When the Wall was built in 1961 three metro lines starting in northern parts of West Berlin, passed through tunnels under the Eastern city centre and ended again in southern parts of West Berlin. The lines concerned were today's underground lines U 6 and U 8 and the S-Bahn line S 2 (today partly also used by other lines). On the sealing off of West Berlin from East Berlin by the Berlin Wall the entrances of the stations on these lines located in East Berlin were shut, however western trains were allowed to continue to pass through without stopping. Passengers in these trains experienced the empty and barely lit ghost stations where time had stood still since 13 August 1961. West Berlin's public transport operator *BVG* (West) paid the east an annual charge in Western Deutsche Marks for its underground lines to use the tunnels under East Berlin. U 6 and S 2 also had one subterranean stop at the Eastern Berlin Friedrichstraße railway station, the only station beneath East Berlin where western U Bahn trains were still allowed to stop. Passengers could change there between U 6, S 2 and the elevated S 3 (then starting and ending in Friedrichstraße) or for the transit trains to West Germany, buy duty free tobacco and liquor for west marks in GDR run kiosks, or enter East Berlin through an inbuilt checkpoint.

Post and telecommunications

West Berlin had its own postal administration first called *Deutsche Post Berlin* (1947–1955) and then Deutsche Bundespost Berlin, separate from West Germany's Deutsche Bundespost, and issuing its own postage stamps until 1990. However the separation was merely symbolic; in reality West Berlin's postal service was completely integrated with West Germany's, using the same postal code system. East and West engaged each other in postal battles in 1948/1949 (during the Blockade) and 1959/1960 (World Year of the Refugees) refusing to transport messages with stamps showing values in the new East or West German currency or with special stamps showing subjects related to the Blockade or the fate of the World War II refugees.

The Post Office also ran the telephone network in Berlin. It was in a sorry state in all four sectors, because by July 1945, before the Western Allies took control of their sectors, the Soviets had dismantled and deported almost all automatic telephone switches, allowing direct dialling instead of operator connected calling. So Berlin's telephone network dropped from hundreds of thousands of connected telephones to a mere 750 in use by end of 1945, all of which were assigned to Allied staff or utility services. Rebuilding the system became a lengthy enterprise because of the post-war economic crisis and the following Berlin Blockade. On 25 February 1946 calls between Berlin and any of the four Allied zones of occupation were again made possible. In April 1949 the Eastern branch of the *Deutsche Post* disconnected all 89 existing telephone lines from West Berlin into the Soviet Zone of occupation in Germany.

Meanwhile West Berlin was integrated into the West German telephone network, using the same international dialling code as West Germany, +49, with the area code 030. On 27 May 1952 the Eastern *Deutsche Post* cut all 4,000 lines connecting East and West Berlin. In order to reduce Eastern tapping of telecommunications between West Berlin and West Germany microwave radio relay connections were built, which wirelessly transmitted telephone calls between antenna towers in West Germany and West Berlin, where two of which were built, one antenna in Berlin-Wannsee and later a second in Berlin-Frohnau, finished on 16 May 1980 with a height of 358 m (1175 ft) (this tower was demolished on 8 February 2009).

Following the détente, on 31 January 1971 East Germany allowed the opening of 10 telephone lines between East and West Berlin. The Western area code for East Berlin was then 00372 (international access prefix 00, East German country code 37, area code 2). Calls from East Berlin were only possible with operator assistance. On 24 June 1972 East Germany opened 32 local exchanges (including Potsdam) in the East German suburbia of West Berlin for calls from West Berlin. From 14 April 1975 East Berliners could once again dial directly to West Berlin, without operator assistance. East Germany conceded to an increase in lines between East and West Berlin to 120 on 15 December 1981. However, private phones were very rare in the East. In 1989, the 17 million East Germans (including East Berliners) were served by only 4 million telephones, only half of which were installed in private homes, the rest being in offices, companies, public telephone kiosks, and the like.

Boroughs of West Berlin

West Berlin comprised the following boroughs:

In the American Sector:

- Neukölln
- Kreuzberg
- Schöneberg
- Steglitz
- Tempelhof
- Zehlendorf

In the British Sector:

- Charlottenburg
- Tiergarten
- Wilmersdorf
- Spandau

In the French Sector:

- Reinickendorf
- Wedding

See also

- 1986 Berlin discotheque bombing
- Berlin Brigade
- Berlin Crisis of 1961
- Bonn
- Ghost station
- History of Germany (1945–1990)
- Judgment in Berlin
- List of Commandants of Berlin Sectors
- Nonviolent revolution
- RAF Gatow
- Spandau Prison
- Stunde Null
- United States Army Berlin

References

[1] The embassies of the Western Allies, later opened in East Berlin, were officially not called, e.g., *Embassy of the French Republic to the German Democratic Republic*, to whose territory East Berlin did not belong according to Western opinion, but *Embassy of the French Republic in Berlin.*

[2] "Germany, East." *Funk & Wagnalls New Encyclopedia,* 11. Funk & Wagnalls, Inc., 1990. ISBN 0-8343-0091-5

[3] According to the German-German *Traffic Agreement* of 29 November 1978, the transit via highway *F 5* was replaced by a new autobahn connecting Hamburg with Wittstock (today's A 24), from there on using the existing autobahn between Berlin and Rostock (today's A 19). The West German *Federal Government* paid DM 1.2 billion to co-finance the construction of these roads. East Germany, in chronic need of western foreign currencies, often showed cooperation whenever Western payments were involved.

[4] This explains the retaining of the name "Deutsche Reichsbahn" despite containing the word "Reich" (Nation or Empire) replaced in the names of all other institutions taken over by the communist GDR.

[5] This was felt in 1980. The Reichsbahn tried to reduce its losses from operating West Berlin's S-Bahns by reducing the staff and the operation time in the evenings and nights, further reducing the salaries of the remaining employees. Being paid worse than West German railway workers, the West Berlin S-Bahn employees went on strike, which was legal in capitalist West Berlin, but illegal in communist East Berlin, because it was regarded as lacking loyalty to the communist party. The strikers occupied the signal towers, blocking any rail traffic in West Berlin as of 20 September. With the help of Soviet patrollers in West Berlin, East German railway workers recaptured the signal towers on 22 September and other railway premises. More than 200 West Berlin Reichsbahn employees who did not return to work were then fired. This was illegal by western law, because striking does not provide legal grounds for a dismissal. However, the Reichsbahn was out of western jurisdiction, so the West Berlin government provided payment of unemployment benefits to the former Reichsbahn workers, despite the Reichsbahn having never paid contributions to unemployment insurance in West Berlin.

[6] East Germany requested East Germans and East Berliners wishing to leave the country to get exit permissions first. However permissions were usually denied, and leaving the country without permission was Republikflucht, considered as a criminal act by the East German justice system.

[7] It took the *BVG* (West) until 1 November 1973 to employ the first female bus driver, by which time all tram lines had been closed down in West Berlin.

[8] *Homecomers* were either German civilians who had been deported into the Soviet Union from those territories it conquered, or former Wehrmacht soldiers and SS fighters, whom the Soviet Union caught as prisoners of war, and all of them from both groups had worked for many years as forced labourers in the Soviet Union, before they were finally allowed to leave.

[9] Literally in English: *Mark of the German Bank of Issue*, which was then the name of the East German state bank.

[10] *Wissenswertes über Berlin: Nachschlagewerk für zuziehende Arbeitnehmer von A-Z* (^{1}1968), Senator für Wirtschaft und Arbeit (ed.), Berlin (West): Senator für Wirtschaft und Arbeit, ^{12}1986, p. 117. No ISBN.

External links

- Berlin 1969 – In the forgotten midpoint of the Cold War (http://www.Berlin1969.com)
- Berlin Exclaves (http://berlin.enclaves.org/)
- History of the Western Allies in Berlin (http://www.western-allies-berlin.com/)

German reunification

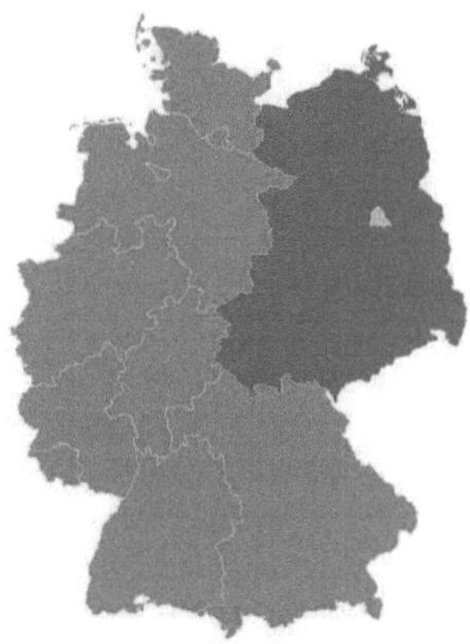

Map showing the division of East and West Germany, with West Berlin in yellow.

German reunification (German: *Deutsche Wiedervereinigung*) was the process in 1990 in which the German Democratic Republic (GDR/East Germany) joined the Federal Republic of Germany (FRG/West Germany), and when Berlin reunited into a single city, as provided by its then Grundgesetz constitution Article 23. The start of this process is commonly referred by Germans as die Wende (*The Turning Point*). The end of the unification process is officially referred to as **German unity** (German: *Deutsche Einheit*), celebrated on 3 October (German Unity Day).[1]

The East German regime started to falter in May 1989, when the removal of Hungary's border fence opened a hole in the Iron Curtain. It caused an exodus of thousands of East Germans fleeing to West Germany and Austria via Hungary. The Peaceful Revolution, a series of protests by East Germans, led to the GDR's first free elections on 18 March 1990, and to the negotiations between the GDR and FRG that culminated in a Unification Treaty,[1] whilst negotiations between the GDR and FRG and the four occupying powers produced the so-called "Two Plus Four Treaty" (*Treaty on the Final Settlement with Respect to Germany*) granting full sovereignty to a unified German state, whose two halves had previously still been bound by a number of limitations stemming from their post-WWII status as occupied regions. The united Germany remained a member of the European Community (later the European Union) and of NATO.

Naming

There is debate as to whether the events of 1990 should be properly referred to as a "reunification" or a "unification". Proponents of the former use the term in contrast with the initial unification of Germany in 1871. Also when the Saarland joined the West German Federal Republic of Germany on 1 January 1957, this was termed the Small Reunification. Popular parlance, which uses "reunification", is deeply affected by the 1989 opening of the Berlin Wall (and the rest of the inner German border) and the physical reunification of the city of Berlin (itself divided only since 1945). Others, however, argue that 1990 represented a "unification" of two German states into a larger entity which, in its resulting form, had never before existed (see History of Germany).

For political and diplomatic reasons, West German politicians carefully avoided the term "reunification" during the run-up to what Germans frequently refer to as *die Wende*. The official[1] and most common term in German is "Deutsche Einheit" (in English "German unity"). German unity is the term that Hans-Dietrich Genscher used in front of international journalists to correct them when they asked him about "reunification" in 1990.

After 1990, the term "die Wende" became more common. The term generally refers to the events (mostly in Eastern Europe) that led up to the actual reunification; in its usual context, this term loosely translates to "the turning point", without any further meaning. When referring to the events surrounding unification, however, it carries the cultural connotation of the time and the events in the GDR that brought about this "turnaround" in German history. However, civil rights activists from Eastern Germany rejected the term *Wende* as it was introduced by SED's Secretary General Egon Krenz.[2]

Precursors to reunification

Further information: Iron_Curtain#Fall_of_the_Iron_Curtain, Revolutions of 1989, Dissolution of the Soviet Union, and European integration

In 1945, the Third Reich ended in defeat and Germany was divided into two separate countries, with the east controlled as part of the Communist Soviet Bloc and the west aligned to Capitalist Europe (which formed into the European Community), including a division in military alliance that formed into the Warsaw Pact and NATO, respectively. The capital city of Berlin was divided into four occupied sectors of control, under the Soviet Union, the United States, the United Kingdom and France. Germans lived under such imposed divisions throughout the ensuing Cold War.

Into the 1980s, the Soviet Union experienced a period of economic and political stagnation, and they correspondingly decreased intervention in Eastern Bloc politics. In 1987, US President Ronald Reagan gave a speech at Brandenburg Gate challenging Soviet leader Mikhail Gorbachev to "tear down this wall" that had separated Berlin. The wall had stood as an icon for the political and economic division between East and West, a division that Churchill had referred to as the "Iron Curtain". In early 1989, under a new era of Soviet policies of *glasnost* (openness), *perestroika* (economic restructuring) and taken to even more progressive levels by Gorbachev, the Solidarity movement took hold in Poland. Further inspired by other images of brave defiance, a wave of revolutions swept throughout the Eastern Bloc that year. In May 1989, Hungary removed their border fence and thousands of East Germans escaped to the West. The turning point in Germany, called "Die Wende", was marked by the "Peaceful Revolution" leading to the removal of the Berlin Wall with East and West Germany subsequently entering into negotiations toward eliminating the division that had been imposed upon Germans more than four decades earlier.

Process of reunification

On 28 November 1989--two weeks after the fall of the Berlin Wall--West German Chancellor Helmut Kohl announced a 10-point program calling for the two Germanies to expand their cooperation with the view toward eventual reunification.[3]

Police officers of the East German Volkspolizei wait for the official opening of the Brandenburg Gate on 22 December 1989.

Initially, no timetable was proposed. However, events rapidly came to a head in early 1990. First, in March, the Party of Democratic Socialism--the former Socialist Unity Party of Germany--was heavily defeated in East Germany's first honest elections. A grand coalition was formed under Lothar de Maizière, leader of the East German wing of Kohl's Christian Democratic Union, on a platform of speedy reunification. Second, East Germany's economy and infrastructure underwent a swift and near-total collapse. While East Germany had long been reckoned as having the most robust economy in the Soviet bloc, the removal of Communist discipline revealed the ramshackle foundations of that system. The East German mark had been practically worthless outside of East Germany for some time before the events of 1989-90, further magnifying the problem.

Discussions immediately began for an emergency merger of the Germanies' economies. On 18 May 1990, the two German states signed a treaty agreeing on monetary, economic and social union. This came into force on 1 July 1990, with the Deutsche Mark replacing the East German mark as the official currency of East Germany. The Deutsche Mark had a very high reputation among the East Germans and was considered stable.[4] While the GDR transferred its financial policy sovereignty to West Germany, the West started granting subsidies for the GDR budget and social security system.[5] At the same time many West German laws came into force in the GDR. This created a suitable framework for a political union by diminishing the huge gap between the two existing political, social, and economic systems.[5]

A reunification treaty between West Germany and the GDR was negotiated in mid-1990, signed on 31 August of that year and finally approved by large majorities in the legislative chambers of both countries on 20 September 1990.[6] After that last step Germany was officially united at 00:00 CET on 3 October 1990. East Germany joined the Federal Republic as the five states of Brandenburg, Mecklenburg-Vorpommern, Saxony, Saxony-Anhalt and Thuringia. These states had been the five original states of East Germany, but had been abolished in 1952 in favour of a centralised system. As part of the 18 May treaty, the five East German states had been reconstituted on 23 August. At the same time, East and West Berlin reunited into one city, which became a city-state along the lines of the existing city-states of Bremen and Hamburg.

The process chosen was one of two options implemented in the West German constitution (Basic Law) of 1949. Via that document's (then-existing) Article 23, any new Länder (German federative states), could adhere to the Basic Law by a simple majority vote. Thus the states' parliaments would vote in for their adhesion. The initial eleven joining states of 1949 comprised the Trizone and West Berlin. However the latter was legally inhibited by Allied objection due to the status of the city as a quadripartite allied occupation area.[7] In 1957 the Saar Protectorate joined West Germany under the Article 23 procedure as Saarland. As the five refounded eastern German states formally joined the Federal Republic using the Article 23 procedure, the area in which the Basic Law was in force simply extended to include them.[8] The alternative would have been for East Germany to join as a whole along the lines of a formal union between two German states that then would have had to, amongst other things, create a new constitution for the newly established country.

Under the model that was chosen, however, the territory of the former German Democratic Republic was simply incorporated into the Federal Republic of Germany, and accordingly the Federal Republic of Germany, now enlarged to include the Eastern States, continued legally to exist under the same legal personality that was founded in May 1949.

Thus, the reunification was not a merger that created a third state out of the two, but an incorporation, by which West Germany absorbed East Germany. Thus, on Unification Day, October 3, 1990, the German Democratic Republic ceased to exist, giving way to five new Federal States, and East and West Berlin were also unified as a single city-state, forming a sixth new Federal State. The new Federal States immediately became parts of the Federal Republic of Germany, so that it was enlarged to include the whole territory of the former East Germany and Berlin.

The practical result of that model is that the now expanded Federal Republic of Germany continued to be a party to all the treaties it had signed prior to the moment of reunification, and thus continued the same membership of the U.N., NATO, the European Communities, etc.; also, the same Basic Law and the same laws that were in force in the Federal Republic continued automatically in force, but now applied to the expanded territory.

To facilitate this process and to reassure other countries, some changes were made to the "Basic Law" (constitution). Article 146 was amended so that Article 23 of the current constitution could be used for reunification. After the five "New Länder" of East Germany had joined, the constitution was amended again to indicate that all parts of Germany are now unified. Article 23 was rewritten and it can still be understood as an invitation to others (e.g. Austria) to join, although the main idea of the change was to calm fears in (for example) Poland, that Germany would later try to rejoin with former eastern territories of Germany that were now Polish or parts of other countries in the East. The changes effectively formalised the Oder-Neisse line as Germany's permanent eastern border. These amendments to the Basic Law were mandated by Article I, section 4 of the Two Plus Four Treaty.[9]

While the Basic Law was modified rather than replaced by a constitution as such, it still permits the adoption of a formal constitution by the German people at some time in the future.

To commemorate the day that marks the official unification of the former East and West Germany in 1990, 3 October has since then been the official German national holiday, the Day of German Unity (*Tag der deutschen Einheit*). It replaced the previous national holiday held in West Germany on 17 June commemorating the Uprising of 1953 in East Germany and the national holiday on 7 October in the GDR.[5]

Foreign opposition

For decades, Germany's allies had stated their support for reunification; Israeli Prime Minister Yitzhak Shamir, who speculated that a country that "decided to kill millions of Jewish people" in the Holocaust "will try to do it again", was one of the few world leaders to publicly oppose it. As reunification became a realistic possibility, however, significant NATO and European opposition emerged in private.[10]

Britain and France

Before the fall of the Berlin Wall, British Prime Minister Margaret Thatcher told Soviet President Mikhail Gorbachev that neither the United Kingdom nor Western Europe wanted the reunification of Germany. Thatcher also clarified that she wanted the Soviet leader to do what he could to stop it, telling Gorbachev "We do not want a united Germany".[11] Although she welcomed East German democracy, Thatcher worried that a rapid reunification might weaken Gorbachev,[12] and favoured Soviet troops staying in East Germany as long as possible to act as a counterweight to a united Germany.[10]

Thatcher, who carried in her handbag a map of Germany's 1937 borders to show others the "German problem", feared that its "national character", size, and central location in Europe would cause the nation to be a "destabilizing rather than a stabilizing force in Europe".[12] In December 1989, she warned fellow European Community leaders at a Strasbourg, France summit Kohl attended, "We defeated the Germans twice! And now they're back!"[13] [10] Although Thatcher had stated her support for German self-determination in 1985,[12] she now argued that Germany's allies had only supported reunification because they had not believed it would ever happen.[10] Thatcher favoured a transition period of five years for reunification, during which the two Germanys would remain separate states. Although she gradually softened her opposition, as late as March 1990 Thatcher summoned historians and diplomats to a seminar at Chequers[12] to ask "How dangerous are the Germans?"[13] and the French ambassador in London reported that Thatcher had told him, "France and Great Britain should pull together today in the face of the German threat."[14] [15]

Similarly, a representative of French President François Mitterrand reportedly told an aide to Gorbachev, "France by no means wants German reunification, although it realises that in the end it is inevitable."[11] At the Strasbourg summit, Mitterrand and Thatcher discussed the fluidity of Germany's historical borders.[10] On 20 January 1990, Mitterrand told Thatcher that a unified Germany could "make more ground than even Hitler had".[14] He predicted that "bad" Germans would reemerge,[13] who might seek to regain former German territory lost after World War II[12] and would likely dominate Hungary, Poland, and Czechoslovakia, leaving "only Romania and Bulgaria for the rest of us". The two leaders saw no way to prevent reunification, however, as "None of us was going to declare war on Germany".[10] Mitterrand recognized before Thatcher that reunification was inevitable and adjusted his views accordingly; unlike her, he was hopeful that participation in a single currency[12] and other European institutions could control a united Germany. Mitterrand still wanted Thatcher to publicly oppose unification, however, to obtain more concessions from Germany.[13]

The rest of Europe

Other European leaders' opinion of reunification was "icy". Italy's Giulio Andreotti warned against a revival of "pan-Germanism", and the Netherlands' Ruud Lubbers questioned the German right to self-determination. They shared Britain and France's concerns over a return to German militarism and the economic power of a reunified nation. The consensus opinion was that reunification, if it must occur, should not occur until at least 1995 and preferably much later.[10]

The four powers

> The United States — and President George H. W. Bush — recognized that Germany had gone through a long democratic transition. It had been a good friend, it was a member of NATO. Any issues that had existed in 1945, it seemed perfectly reasonable to lay them to rest. For us, the question wasn't should Germany unify? It was how and under what circumstances? We had no concern about a resurgent Germany...
>
> —Condoleezza Rice, Special Assistant to the President for National Security Affairs[16]

The victors of World War II—the United States, the Soviet Union, the United Kingdom, and France, comprising the Four-Power Authorities—retained authority over Berlin, such as control over air travel and its political status. The Soviet Union sought early to use reunification as a way to push Germany out of NATO into neutrality, removing nuclear weapons from its territory. West Germany interpreted a 21 November 1989 diplomatic message on the topic, however, as meaning that only two weeks after the Wall's collapse the Soviet leadership already anticipated reunification. This belief, and the worry that his rival Genscher might act first, encouraged Kohl on November 28 to announce a detailed reunification plan (10-Point Plan). While his speech was very popular within West Germany Kohl's speech caused concern among other European governments, with whom he had not discussed the plan.[10]

The Americans did not share the Europeans' and Russians' historical fears over German expansionism, but wished to ensure that Germany would stay within NATO. In December 1989, the administration of President George H. W. Bush made a united Germany's continued NATO membership a requirement for supporting reunification. Kohl agreed, although less than 20% of West Germans supported remaining within NATO; he also wished to avoid a neutral Germany, as he believed that would destroy NATO, cause the United States and Canada to leave Europe, and Britain and France would form an alliance. The United States increased its support of Kohl's policies, as it feared that otherwise Oskar Lafontaine, a critic of NATO, might become Chancellor.[10]

Helmut Kohl became first chancellor of a reunified Germany.

Horst Teltschik, Kohl's foreign policy advisor, later recalled that Germany would have paid "100 billion deutschmarks" had the Soviets demanded it. The USSR did not make such great demands, however, with Gorbachev stating in February 1990 that "The Germans must decide for themselves what path they choose to follow". In May 1990 he repeated his remark in the context of NATO membership while meeting Bush, amazing both the Americans and Germans.[10]

Conclusion

During a NATO-Warsaw Pact conference in Ottawa, Canada, Genscher persuaded the four powers to treat the two Germanys as equals instead of defeated junior partners, and for the six nations to negotiate alone. Although the Dutch, Italians, Spanish, and other NATO powers opposed such a structure, which meant that the alliance's boundaries would change without their participation, the six nations began negotiations in March 1990. After Gorbachev's May agreement on German NATO membership, the Soviets further agreed that Germany would be treated as an ordinary NATO country, with the exception that former East German territory would not have foreign NATO troops or nuclear weapons. In exchange, Kohl agreed to reduce the sizes of both Germanys' militaries, renounce weapons of mass destruction, and accept the postwar Oder-Neisse line as

Germany's eastern border. In addition, Germany agreed to pay about 55 billion deutschmarks to the Soviet Union in gifts and loans, the equivalent of eight days of the West German GDP.[10]

Mitterrand agreed to reunification in exchange for a commitment from Kohl to the European Economic and Monetary Union.[17] The British insisted to the end, against Soviet opposition, that NATO be allowed to hold maneuvers in the former East Germany. After the Americans intervened,[10] both the UK and France ratified the Treaty on the Final Settlement with Respect to Germany in September 1990, thus finalizing the reunification for purposes of international law. Thatcher later wrote that her opposition to reunification had been an "unambiguous failure".[12]

Aftermath

On 14 November 1990, the united Germany and Poland signed the German–Polish Border Treaty, finalizing Germany's boundaries as permanent along the Oder-Neisse line, and thus, renouncing any claims to Silesia, East Brandenburg, Farther Pomerania, and territories of the former province of East Prussia.[18] The treaty also granted certain rights for political minorities on either side of the border.[19] The following month, the first all-German free elections since 1932 were held, resulting in an increased majority for the coalition government of Chancellor Helmut Kohl.

On 15 March 1991, the Treaty on the Final Settlement With Respect to Germany entered into force, putting an end to the remaining limitations on German sovereignty that resulted from the post WWII arrangements.

Inner reunification

Vast differences between the former East Germany and West Germany (for example, in lifestyle, wealth, political beliefs and other matters) remain, and it is therefore still common to speak of eastern and western Germany distinctly. The eastern German economy has struggled since unification, and large subsidies are still transferred from west to east. The former East Germany area has often been compared to the underdeveloped Southern Italy and the Southern United States during Reconstruction after the American Civil War. While the East German economy has recovered recently, the differences between East and West remain constant.[20] [21]

Reenacting noted Alfred Eisenstaedt V-J Day photo at 12:01 A.M. on Reunification night in Cologne, Germany.

Politicians and scholars have frequently called for a process of "inner reunification" of the two countries and asked whether there is "inner unification or continued separation".[22] "The process of German unity has not ended yet", proclaimed Chancellor Angela Merkel, who grew up in East Germany, in 2009.[23] Nevertheless, the question of this "inner reunification" has been widely discussed in the German public, politically, economically, culturally, and also constitutionally since 1989.

Politically, since the fall of the Wall, the successor party of the former East German socialist state party has become a major force in German politics. It was renamed PDS, and, later, merged with the Western leftist party WASG to form the party Die Linke (The Left).[24]

Constitutionally, the Basic Law (Grundgesetz), the West German constitution, provided two pathways for a unification, the first including an implementation of the constitution in the East, the second the implementation of a new all-German constitution, safeguarded by a popular referendum. While the former option was chosen as the most feasible one, the latter option was partly regarded as a means to foster the "inner reunification".[25] [26]

A public manifestation of coming to terms with the past (*Vergangenheitsbewältigung*) is the existence of the so-called Birthler-Behörde, the Federal Commissioner for the Stasi Archives, which collects and maintains the files of the East German security apparatus.[27]

Berlin Wall at the Brandenburg Gate, 10 November 1989

The economic reconstruction of former Eastern Germany following the reunification required large amounts of public funding which turned some areas into boom regions, although overall unemployment remains higher than in the former West.[28] Reunification did, however, lead to a large rise in the average standard of living in former Eastern Germany. Between 1990 and 1995, gross wages in the east rose from 35% to 74% of western levels, while pensions rose from 40% to 79%.[29] :209

In terms of media usage and reception, the country remains partially divided especially among the older generations. Mentality gaps between East and West persist, but so does sympathy.[23] Additionally, the daily-life exchange of Easterners and Westerners is not so large as expected.[30] [31] Young people's knowledge of the former East Germany is very low.[32] Some people in Eastern Germany engage in "Ostalgia", which is a certain nostalgia for the time before the wall came down.[33]

Placard found in all roads between Western and Eastern Germany that were blocked during division. Text reads: "Here, Germany and Europe were divided until December 10, 1989 at 10:15 AM". The date and time vary according to the actual moment when a particular crossing was opened.

Today, there are several prominent East Germans shaping the image of Germany at home and abroad, including Kurt Masur, Michael Ballack, Katarina Witt, and Angela Merkel.

See also

- History of Germany since 1945
- Reunification
- Stalin Note – 1952 German reunification proposal
- Transitology
- Goodbye Lenin!

References

[1] Vertrag zwischen der Bundesrepublik Deutschland und der Deutschen Demokratischen Republik über die Herstellung der Einheit Deutschlands (Einigungsvertrag) (http://bundesrecht.juris.de/einigvtr/BJNR208890990.html) Unification Treaty signed by the Federal Republic of Germany and the German Democratic Republic in Berlin on 31 August 1990 (official text, in German).

[2] Krenz speech of October 18, 1989 in the German radio archive (http://1989.dra.de/ton-und-videoarchiv/archivnachweise.html?tx_weeaardra_pi2[what]=031846,031846&tx_weeaardra_pi2[from]=38&tx_weeaardra_pi2[ids]=302,146&tx_weeaardra_pi2[backId]=38&cHash=ba89eab5f1)

[3] Helmut Kohl's 10-Point Plan for German Unity (http://germanhistorydocs.ghi-dc.org/sub_document.cfm?document_id=223)

[4] German Unification Monetary union (http://www.cepr.org/Pubs/bulletin/dps/dp719.htm). Cepr.org (1990-07-01). Retrieved on 2010-10-19.

[5] Embassy of the Federal Republic of Germany London – A short history of German reunification (http://www.london.diplo.de/Vertretung/london/en/01/Feste/Tag__der__Deutschen__Einheit/History__of__reunification__seite.html). London.diplo.de. Retrieved on 2010-10-19.

[6] Opening of the Berlin Wall and Unification: German History (http://www.germanculture.com.ua/library/history/bl_wall_collapse_unification.htm). Germanculture.com.ua. Retrieved on 2010-10-19.

[7] However, in many fields it functioned de facto as if it were a component State of West Germany.

[8] Germany Today – The German Unification Treaty – travel and tourist information, flight reservations, travel bargains, hotels, resorts, car hire (http://europe-today.com/germany/gerunif.html). Europe-today.com. Retrieved on 2010-10-19.
[9] "File:Zwei-Plus-Vier-Vertrag.pdf" (http://commons.wikimedia.org/wiki/File:Zwei-Plus-Vier-Vertrag.pdf). .
[10] Wiegrefe, Klaus (2010-09-29). "An Inside Look at the Reunification Negotiations" (http://www.spiegel.de/international/germany/0,1518,druck-719848,00.html). *Der Spiegel*. . Retrieved 2010-10-04.
[11] "Thatcher told Gorbachev Britain did not want German reunification" (http://www.timesonline.co.uk/tol/news/politics/article6829735.ece). *Michael Binyon* (London: Times). 11 September 2009. . Retrieved 8 November 2009.
[12] Kundnani, Hans (2009-10-28). "Margaret Thatcher's German war" (http://entertainment.timesonline.co.uk/tol/arts_and_entertainment/the_tls/article6893915.ece). *The Times*. . Retrieved 2010-10-05.
[13] Volkery, Carsten (2009-11-09). "The Iron Lady's Views on German Reunification/'The Germans Are Back!'" (http://www.spiegel.de/international/europe/0,1518,648364,00.html). *Der Spiegel*. . Retrieved 2010-10-05.
[14] Anne-Laure Mondesert (AFP) (31 October 2009). "London and Paris were shocked by German reunification" (http://www.calgaryherald.com/news/London+Paris+were+shocked+German+reunification/2168902/story.html). *Calgary Herald*. . Retrieved 9 November 2009.
[15] Peter Allen (2 November 2009). "Margaret Thatcher was 'horrified' by the prospect of a reunited Germany" (http://www.telegraph.co.uk/news/worldnews/europe/germany/6480413/Margaret-Thatcher-was-horrified-by-the-prospect-of-a-reunited-Germany.html). *Telegraph* (London). . Retrieved 9 November 2009.
[16] "'I Preferred To See It as an Acquisition'" (http://www.spiegel.de/international/world/0,1518,719444,00.html). Der Spiegel. September 29, 2010. . Retrieved 2010-10-07.
[17] Ben Knight (8 November 2009). "Germany's neighbors try to redeem their 1989 negativity" (http://www.dw-world.de/dw/article/0,,4861759,00.html). *Deutsche Welle*. . Retrieved 9 November 2009.
[18] The territory of the League of Nations mandate of the Free City of Danzig, annexed by Poland in 1945 and comprising the city of Gdańsk (Danzig) and a number of neighbouring cities and municipalities, had never been claimed by any official side, because West Germany followed the legal position of Germany in its borders of 1937, thus before any Nazi annexations.
[19] Poland Germany border - Oder Neisse (http://polandpoland.com/germany_poland_border.html). Polandpoland.com (1990-11-14). Retrieved on 2010-10-19.
[20] Underestimating East Germany - Magazine (http://www.theatlantic.com/doc/200911u/east-german-economy). The Atlantic (2010-10-15). Retrieved on 2010-10-19.
[21] After the fall 20 years ago this week the crumbling of the Berlin Wall began an empire s end (http://www.annistonstar.com/pages/full_story/push?article-After+the+fall-+20+years+ago+this+week-+the+crumbling+of+the+Berlin+Wall+began+an+empire-s+end&id=4380443-After+the+fall-+20+years+ago+this+week-+the+crumbling+of+the+Berlin+Wall+began+an+empire-s+end&instance=home_opinion). Anniston Star. Retrieved on 2010-10-19.
[22] National identity in eastern Germany ... - Google Bücher (http://books.google.com/books?id=LN6dpmV0p40C&dq=Inner+reunification&hl=de&source=gbs_navlinks_s). Books.google.com (1990-10-30). Retrieved on 2010-10-19.
[23] (German) Umfrage: Ost- und Westdeutsche entfernen sich voneinander - Nachrichten Politik - WELT ONLINE (http://www.welt.de/politik/article3775359/Ost-und-Westdeutsche-entfernen-sich-voneinander.html). Welt.de. Retrieved on 2010-10-19.
[24] (German) Partei des Demokratischen Sozialismus – Wikipedia. De.wikipedia.org. Retrieved on 2010-10-19.
[25] Gastbeitrag: Nicht für die Ewigkeit - Staat und Recht - Politik (http://www.faz.net/s/RubD5CB2DA481C04D05AA471FA88471AEF0/Doc~E4120761F98224225BEB5F94E7F91077D~ATpl~Ecommon~Scontent.html). Faz.Net. Retrieved on 2010-10-19.
[26] Aus Politik und Zeitgeschichte, Nr. 18 2009, 27.04.2009 - Das Grundgesetz - eine Verfassung auf Abruf? (http://www.das-parlament.de/2009/18-19/Beilage/004.html). Das-parlament.de. Retrieved on 2010-10-19.
[27] DDR-Geschichte: Merkel will Birthler-Behörde noch lange erhalten - SPIEGEL ONLINE - Nachrichten - Politik (http://www.spiegel.de/politik/deutschland/0,1518,601549,00.html). Spiegel.de (2009-01-15). Retrieved on 2010-10-19.
[28] Facts about Germany: Society (http://www.tatsachen-ueber-deutschland.de/en/society.html). Tatsachen-ueber-deutschland.de. Retrieved on 2010-10-19.
[29] Parkes, K. Stuart (1997). *Understanding contemporary Germany* (http://books.google.com/books?id=htUiBNqUAnYC&lpg=PP1&dq=understanding contemporary germany parkes&pg=PA209#v=onepage&q=74%&f=false). Taylor & Francis. ISBN 0415141249. .
[30] (German) Ostdeutschland: Das verschmähte Paradies | Campus | ZEIT ONLINE (http://www.zeit.de/campus/2008/05/ost-west-unis). Zeit.de (2008-09-29). Retrieved on 2010-10-19.
[31] (German) Partnerschaft: Der Mythos von den Ost-West-Ehepaaren - Nachrichten Panorama - WELT ONLINE (http://www.welt.de/vermischtes/article3820575/Der-Mythos-von-den-Ost-West-Ehepaaren.html). Welt.de. Retrieved on 2010-10-19.
[32] Politics and History - German-German History - Goethe-Institut (http://www.goethe.de/ges/pok/ddg/en3822678.htm). Goethe.de. Retrieved on 2010-10-19.
[33] Zeitchik, Steven (7 October 2003). "German Ostalgie : Fondly recalling the bad old days" (http://www.nytimes.com/2003/10/07/opinion/07iht-edzeitchik_ed3_.html?pagewanted=). *The New York Times*. . Retrieved 3 May 2010.

Further reading

- Charles S. Maier, *Dissolution: The Crisis of Communism and the End of East Germany* (Princeton University Press, 1997).
- Philip Zelikow and Condoleezza Rice, *Germany Unified and Europe Transformed: A Study in Statecraft* (Harvard University Press, 1995 & 1997).

Primary sources

- Jarausch, Konrad H., and Volker Gransow, eds. *Uniting Germany: Documents and Debates, 1944-1993* (1994), primary sources in English translation

External links

- Germany's Eastern Burden: The Price of a Failed Reunification (http://www.spiegel.de/international/spiegel/0,1518,373639,00.html)
- The End of East Germany (http://www.dw-world.de/dw/article/0,1564,1687204,00.html)
- German Embassy Publication, Infocus: German Unity Day (http://timpani.globescope.com/relaunch/info/publications/infocus/15yrs_Reunification/index.html)
- The Unification Treaty (Berlin, 31 August 1990) (http://www.ena.lu/europe/market-amsterdam/unification-treaty-berlin-1990.htm) website of European navigator
- An Inside Look at the Reunification Negotiations (http://www.spiegel.de/international/germany/0,1518,719848,00.html)

East Germany

German Democratic Republic
Deutsche Demokratische Republik

←
 **1949–1990**  →
←

Flag

Coat of arms

Motto
"Proletarier aller Länder, vereinigt Euch!"
("Workers of all nations, unite !")

Anthem
"Auferstanden aus Ruinen"
("Risen from Ruins")

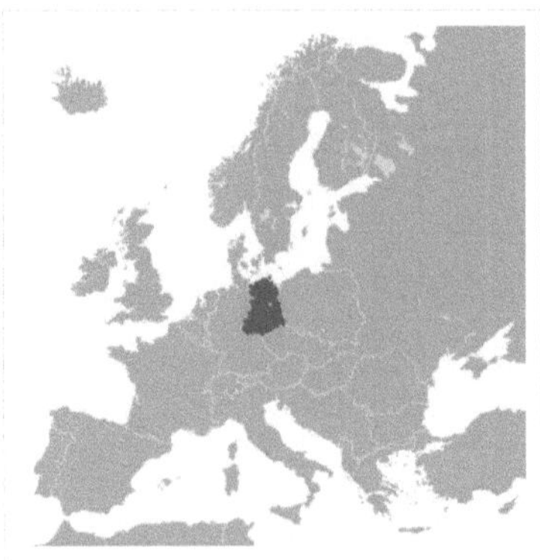

Capital	East Berlin
Language(s)	Official: German Unofficial minority languages: Sorbian
Government	Socialist republic
Chairman of the Council of State	
- 1949–60	Wilhelm Pieck (*President*)
- 1960–73	Walter Ulbricht
- 1973–76	Willi Stoph

- 1976–89	Erich Honecker
- 1989	Egon Krenz
- 1989–90	Manfred Gerlach
Chairman of the Council of Ministers	
- 1949–64	Otto Grotewohl
- 1964–73	Willi Stoph
- 1973–76	Horst Sindermann
- 1976–89	Willi Stoph
- 1989–90	Hans Modrow
- 1990	Lothar de Maizière
Legislature	Volkskammer
Historical era	Cold War
- Established	7 October 1949
- Final settlement	25 September 1990
- German reunification	3 October 1990
Area	
- 1990	108333 km^2 (41828 sq mi)
Population	
- 1990 est.	16111000
Density	148.7 /km^2 (385.2 /sq mi)
Currency	Mark der DDR (M) until 30 June 1990, named: 1948–64 Deutsche Mark (DM) 1964–67 Mark der Deutschen Notenbank (MDN) Deutsche Mark (DM) as of 1 July 1990
Internet TLD	.dd[1]
Calling code	+37
Today part of	Germany
[1] Although .dd was reserved as corresponding ISO code for East Germany, it was not entered to the root before the country was disestablished.[1]	
[2] Country code +37 was withdrawn in 1992; the numbers range was divided into ten new country codes, re-allocated among several post-Soviet states and European microstates.	

The **German Democratic Republic** (**GDR**; German: **Deutsche Demokratische Republik** [ˈdɔʏtʃə demoˈkʀaːtɪʃə ʀepuˈbliːk] or **DDR**), informally called **East Germany** (German: **Ost-Deutschland**) by West Germany and other countries, was a socialist state established in 1949 in the Soviet zone of occupied Germany, including East Berlin of the Allied-occupied capital city. The German Democratic Republic, which consisted geographically of post-WWII northeast Germany rather than all of eastern Germany, had an area of 107,771 km^2 (41,610 mi^2), bordering Czechoslovakia in the south, West Germany (officially: Federal Republic of Germany) in the south and west, the Baltic Sea to the north, and Poland in the east.

In 1989 a non-violent revolution overthrew the Communists. The Soviet Union refused to intervene and the country soon reunited with West Germany, and is now part of Germany.

At German reunification on October 3, 1990, new *Länder* (states) were formed within the area occupied by the former German Democratic Republic. Following historical borders of provinces that existed before 1952, new federal states (in German usually referred to as *Die Neuen Bundesländer*; Engl. new federal states) were formed and integrated into the Federal Republic of Germany (FRG). Moreover, the German Democratic Republic was disestablished after the Communist government, of the Socialist Unity Party of Germany (SED), lost the general election on March 18, 1990, and thus its parliamentary majority in the *Volkskammer* (People's Chamber); subsequently, on August 23, 1990, the *Volkskammer* re-established the five pre-war states — Brandenburg, Mecklenburg-Vorpommern, Saxony, Saxony-Anhalt, and Thuringia (disestablished in 1952) — for the reunification of East Germany to West Germany.

Naming conventions

The official name was *Deutsche Demokratische Republik* (German Democratic Republic), usually abbreviated as *DDR*. Both terms were used in East Germany with an increasing emphasis on the abbreviated name, especially since East Germany considered West Germans and West Berliners to be foreigners following the promulgation of its second constitution in 1968.

Ostzone (Eastern Zone) or *Soviet Zone* were two surrogate names for East Germany that were often used colloquially. The different names used to describe the German Democratic Republic reflected political positions during the Cold War conflict; for example, many Westerners doubted the political sovereignty and democratic constitution of East Germany. Surrogate name usage for East Germany could thus reveal the political leaning of a person or news source. So the media controlled by the East German government emphasised the use of the official name, DDR, while West Germans, western media and statesmen may have used other names such as Middle Germany, emphasising the location of East Germany in the centre of pre-1937 Germany.

The name Sowjetische Besatzungszone (Soviet Occupation Zone, often abbreviated to SBZ) was used by those who wanted to indicate that East Germany lacked sovereignty, whereas others used *Ostzone* or *der Osten* (English: *Eastern Zone* or *the East*) to avoid the actual name of the state. The latter term, because it was based plainly on geographic location, was sometimes also used by East Germans. Some West German media referred to East Germany initially as the *SBZ* and later consistently named it the *so-called "GDR"* (sogenannte "DDR").[2]

However, over time East Germany's abbreviation *DDR* became colloquial also among most West Germans and West German media.[3] *Ostdeutschland* (an ambiguous term meaning simultaneously East or Eastern Germany) was not commonly used in East or West German common parlance to refer to the German Democratic Republic, because *Ostdeutschland* usually referred to the Former eastern territories of Germany.

The term *Westdeutschland* (West Germany) when used by West Germans was almost always a reference to the geographic region of Western Germany but not to the area within the boundaries of the Federal Republic of Germany. However, this usage was not always consistent, as, for example, West Berliners frequently applied the term *Westdeutschland* to denote the Federal Republic.

History

Further information: History of Germany

Explaining the internal impact of the DDR regime from the perspective of German history in the long-term, historian Gerhard A. Ritter (2002) has argued that the East German state was defined by two dominant forces – Soviet Communism on the one hand, and German traditions filtered through the interwar experiences of German Communists on the other. It always was constrained by the powerful example of the increasingly prosperous West, to which East Germans compared their nation. The changes wrought by the Communists were most apparent in ending capitalism and transforming industry, agriculture, in the militarization the society, and the political thrust of the educational system and the media. On the other hand, there was relatively little change made in the historically independent domains of the sciences, the engineering professions, the Protestant churches, and in many bourgeois life styles. Social policy, says Ritter, became a critical legitimization tool in the last decades and mixed Communist and traditional elements about equally.[4]

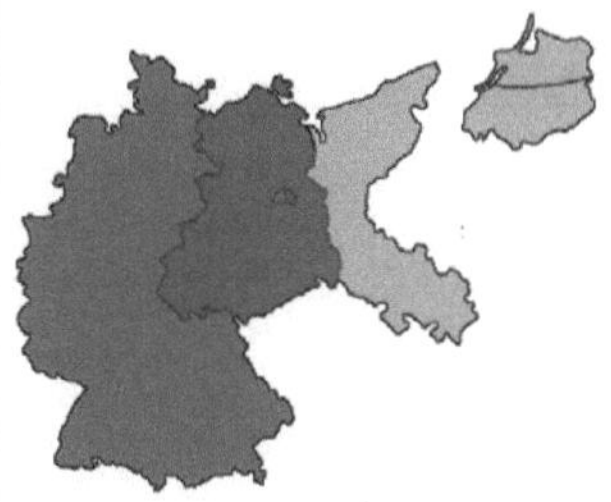

Germany defeated: Per the Potsdam Conference, the Allies jointly occupied Germany west of the Oder-Neisse line.

Origins

At the Yalta Conference during World War II, the Allies (U.S., Britain, and the Soviet Union) agreed on dividing a defeated Germany into occupation zones,[5] and on dividing Berlin, the German capital, among the Allied powers as well. Initially this meant the construction of three zones of occupation, i.e. American, British, and Soviet. Later, a French zone was carved out of the American and British zones.

GDR created 1949

The ruling Communist party, known as the "Socialist Unity Party" (SED), was formed in April 1946 out of the forced merger between the German Communist Party (KPD) and the Social Democratic Party of Germany (SPD). As Walter Ulbricht noted, everything was made to look democratic while in reality Communists retained control in the background. They were totally loyal to Stalin, and realized their regime would collapse if it lost the backing of the Soviet army—as indeed happened in 1989. Historians debate whether the decision to form a separate country was initiated by Stalin or by the SED.[6]

As West Germany was reorganized and gained independence from the occupation, Stalin established the German Democratic Republic in 1949. The creation of the two states made the 1945 division of Germany permanent.[7]

In 1949 the Soviets turned control of East Germany over to the Communist Party, headed by Wilhelm Pieck (1876–1960), who became president of the GDR and remained officially 'Number One' until his death in 1960, while any real power allowed by the Soviets was assumed by SED General Secretary Walter Ulbricht. The old Socialist Party was taken over by the Communists, and Socialist leader Otto Grotewohl (1894–1964) became prime minister.

West Germany saw itself as the legal successor to the Third Reich, shouldering the burdens of legal responsibility for its crimes. By contrast, East Germany renounced ties to the Nazi past, styling itself the "anti-fascist rampart" and proclaiming itself the first socialist state on German soil. It refused to admit the existence of anti-semitism and refused to recognize Israel or reimburse victims of the Holocaust.[8]

Soviet role

In 1955, the USSR declared the Soviet occupation zone — the historic middle portion of Germany — to be a sovereign state named the *Deutsche Demokratische Republik* (German Democratic Republic, established in 1949), while the Red Army and the Western Allies' occupation forces remained in place under the tripartite Potsdam Agreement (1945) which established the Allied Occupation of Germany.

The Communist German Democratic Republic was established in the historic "Mitteldeutschland" (Middle Germany). Former German territories east of the Oder and Neisse rivers, mainly the Prussian provinces of Pomerania, East Prussia, West Prussia, Upper Silesia, Lower Silesia, and the eastern Neumark of Brandenburg were thus detached from Germany. To compensate Poland for the USSR's annexation of its eastern provinces, the Allies provisionally established Poland's post-war western border at the Oder-Neisse line at the Yalta Conference (1945). As a result, most of Germany's central territories became the *Sowjetische Besatzungszone* (SBZ, Soviet Occupation Zone). All other lands east of the Oder–Neisse line were put under Polish administration, with the exception of historic northern East Prussia, which went to the USSR.

Zones of occupation

Further information: Allied-occupied Germany

In the Yalta and Potsdam conferences, the Allies established their joint military occupation and administration of Germany via the Allied Control Council (ACC), a four-power (US, UK, USSR, France) military government effective until the restoration of German sovereignty. In eastern Germany, the Soviet Occupation Zone (SBZ – *Sowjetische Besatzungszone*) comprised the five states (*Länder*) of Mecklenburg-Vorpommern, Brandenburg, Saxony, Saxony-Anhalt, and Thuringia. Disagreements over the policies to be followed in the occupied zones quickly led to a breakdown in cooperation between the four powers, and the Soviets administered their zone without regard to the policies implemented in the other zones. The Soviets withdrew from the ACC in 1948; subsequently as the other three zones were increasingly unified and granted self-government, the Soviet administration facilitated the development of a separate socialist government in its zone.

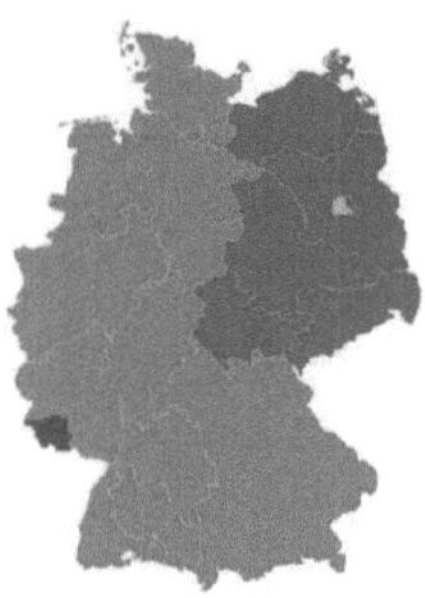

Germany 1949: West Germany (blue) comprised the Western Allies' zones, excluding the Saarland (purple); the Soviet zone, East Germany (red) surrounded West Berlin (yellow)

Yet, seven years after the Allies' Potsdam Agreement to a unified Germany, the USSR via the Stalin Note (March 10, 1952) proposed German reunification and superpower disengagement from Central Europe, which the three Western Allies (US, France, UK) rejected. Soviet leader Joseph Stalin, a communist proponent of reunification, was dead by March 1953. Similarly, Lavrenty Beria, the First Deputy Prime Minister of the USSR, pursued German reunification, but an internal (Party) *coup d'étât* deposed him from government in mid-1953, before he could act on the matter. His successor, Nikita Khrushchev, rejected reunification as equivalent to returning East Germany for annexation to the West; hence reunification went unconsidered until the German Democratic Republic collapsed in 1989.

East Germany and the Eastern Bloc diplomatically recognised East Berlin as the capital city of the German Democratic Republic, but the Western Allies disputed said recognition, considering the entire city of Berlin an occupied territory governed by the martial law of the Allied Control Council. According to Margarete Feinstein, East Berlin's status as the capital was largely unrecognized by the West and most Third World countries. [9] In practice, the ACC's authority was rendered moot by the Cold War, and the East German government ignored the legal restrictions on integration of East Berlin into the GDR.

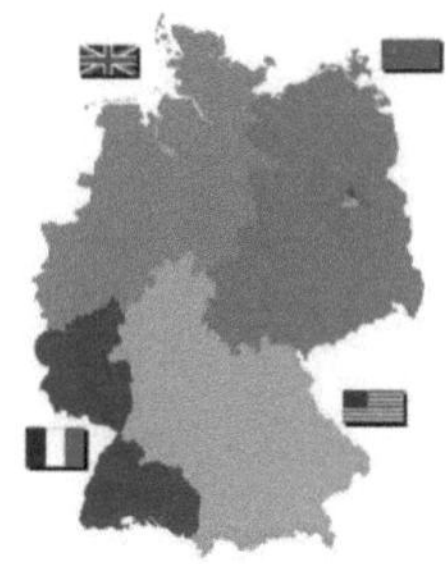

Post-war occupied Germany: The British (green), Soviet (Red), American (Yellow), and French (Blue) occupation zones.

Abetted by ACC's weakness, Cold War political conflicts among the Allies over the status of West Berlin provoked the Berlin Blockade (June 24, 1948 – May 12, 1949), in which the Soviet army stopped all Allied rail, road, and water traffic to and from West Berlin. The Allies countered the Soviets with the Berlin Airlift (1948–49) of food, fuel, and supplies to keep West Berlin alive.[10]

Partition

In 1946, the Communist Party of Germany (*Kommunistische Partei Deutschlands* — KPD) and the Social Democratic Party of Germany (*Sozialdemokratische Partei Deutschlands* — SPD) merged to form the Socialist Unity Party of Germany (SED — *Sozialistische Einheitspartei Deutschlands*) (April 21, 1945), which then won the elections of 1946, held under the oversight of the Soviet army. Being a Marxist-Leninist political party, the SED's government nationalised infrastructure and industrial plants.

GDR leaders: President Wilhelm Pieck and Prime Minister Otto Grotewohl, 1949

In 1948, the German Economic Commission (Deutsche Wirtschaftskomission—DWK) under its chairman Heinrich Rau assumed administrative authority in the Soviet occupation zone, thus becoming the predecessor of an East German government.[11] [12]

On October 7, 1949, the SED established the *Deutsche Demokratische Republik* (German Democratic Republic — GDR), based on a socialist political constitution establishing its control of the anti-fascist National Front of the German Democratic Republic (NF — *Nationale Front der Deutschen Demokratischen Republik*), an omnibus alliance of every party and mass organisation in East Germany. The NF was established to stand for election to the *Volkskammer* ("People's Chamber"), the East German parliament. The first (and only) President of the German Democratic Republic was Wilhelm Pieck. However, after 1950, the true ruler of East Germany was Walter Ulbricht, the First Secretary of the SED.[13]

SED First Secretary, Walter Ulbricht, 1950

On June 16, 1953, workers constructing the new *Stalinallee* boulevard in East Berlin rioted against a 10% production quota increase. Initially a labour protest, it soon included the general populace, who added their anti-Soviet discontent to the workers' civil disobedience, and on June 17 similar protests occurred throughout the GDR, with more than a million people striking in some 700 cities and towns. Fearing anti-communist counter-revolution on June 18, 1953, the government of the GDR enlisted the Soviet Occupation Forces to aid the *Volkspolizei* ("People's Police") in suppressing the rioters; some fifty people were killed and some 10,000 were jailed.[14] [15] (See Uprising of 1953 in East Germany.)

The German war reparations owed to the USSR impoverished the Soviet Zone of Occupation and severely weakened the East German economy. In the 1945–46 period, the Soviets confiscated and transported to the USSR approximately 33% of the industrial plant and by the early 1950s had extracted some 10 billion dollars in reparations in agricultural and industrial products.[16]

The poverty of East Germany induced by reparations provoked the *Republikflucht* ("flight from the republic") to West Germany, aggravating the emigration, continual since the 1940s, from the Soviet zone of Germany to the Western Allied zones, further weakening the GDR's economy. Western economic opportunities and lack of political freedom in East Germany induced a brain drain. In response, the GDR closed the Inner German Border, and on the night of August 12–13, 1961, East German soldiers began erecting the Berlin Wall which prevented anyone from escaping.[17]

Head of State: Erich Honecker (1971–89)

In 1971, Soviet leader Leonid Brezhnev had Ulbricht removed; Erich Honecker replaced him. While the Ulbricht government had experimented with liberal reform, the Honecker government increased controls upon the populace of the GDR. The new government introduced a new East German Constitution which defined the German Democratic Republic as a "republic of workers and peasants" and hardly mentioned the word "German".[18]

Initially, East Germany maintained that it was the only lawful government of Germany. However, from the 1960s onward, East Germany held itself out as a separate country from West Germany, and shared the legacy of the united German state of 1871–1945. West Germany, in contrast, claimed an exclusive mandate for all of Germany. From 1949 to the early 1970s, West Germany maintained that East Germany was an illegally constituted state. It argued that the GDR was a Soviet puppet regime and thus illegitimate. This position was shared by most of the world, until 1973. East Germany was recognized only by Communist countries and the Arab bloc, along with some "scattered sympathizers".[19] According to the Hallstein Doctrine (1955), West Germany also did not diplomatically recognize any country — except the USSR — that recognized East German sovereignty.

But in the early 1970s, the *Ostpolitik* ("Eastern Policy") of "Change Through Rapprochement" of the pragmatic government of FRG Chancellor Willy Brandt, established normal diplomatic relations with the East Bloc states and the GDR. In the event, the Treaty of Moscow (August 1970), the Treaty of Warsaw (December 1970), the Four Power Agreement on Berlin (September 1971), the Transit Agreement (May 1972), and the Basic Treaty (December 1972) established normal relations between the Germanies, later allowing their integration to the United Nations. This also increased the number of countries recognizing East Germany to 55, including the US, UK and France, though the last three still refused to recognize East Berlin as the capital, and insisted on a specific provision in the

UN resolution accepting the two Germanies into the UN to that affect.[19]

GDR identity

From the beginning, the newly formed GDR tried to establish its own separate identity. Because of Marx's abhorrence of Prussia, the SED repudiated continuity between Prussia and the GDR. The SED destroyed the Junker manor houses, wrecked the Berlin city palace and removed the equestrian statue of Frederick the Great from East Berlin. Instead the SED focused on the progressive heritage of German history, including Thomas Müntzer's role in the German Peasants' War and the role played by the heroes of the class struggle during Prussia's industrialization. Nevertheless, as early as 1956 East Germany's Prussian heritage asserted itself in the NVA.

GDR era Karl Marx monument in Chemnitz (renamed *Karl-Marx-Stadt* from 1953 to 1990).

As a result of the Ninth Party Congress in May 1976, East Germany after 1976–77 considered its own history as the essence of German history, in which West Germany was only an episode. It laid claim to reformers such as Karl Freiherr vom Stein, Karl August von Hardenberg, Wilhelm von Humboldt, and Gerhard von Scharnhorst.

In the early 1980s, West Germany adopted the line of "two German states in one German nation". While it respected East Germany's independence, it formally maintained that the GDR was merely a *de facto* government within a single German nation of which the FRG was the sole representative. For instance, it did not treat East Germans as foreigners.

The *Wende*

Police cadets of the East German Volkspolizei wait for the official opening of the Brandenburg Gate on 22 December 1989.

In 1989, following widespread public anger over the results of local government elections that spring, many citizens applied for exit visas or left the country illegally. In August 1989 Hungary removed its border restrictions and unsealed its border, and more than 13,000 people left East Germany by crossing the "green" border via Czechoslovakia into Hungary and then on to Austria and West Germany.[20] Many others demonstrated against the ruling party, especially in the city of Leipzig. Kurt Masur, the conductor of the Leipzig Gewandhaus Orchestra, led local negotiations with the government and held town meetings in the concert hall.[21] The demonstrations eventually led Erich Honecker to resign in October, and he was replaced by a slightly more moderate communist, Egon Krenz.

On November 9, 1989, a few sections of the Berlin Wall were opened, resulting in thousands of East Germans crossing into West Berlin and West Germany for the first time. Krenz resigned a few days later, and the SED abandoned power shortly afterward. Although there were some limited attempts to create a permanent democratic East Germany, these were soon overwhelmed by calls for unification with West Germany.

East Germany held its last elections in March 1990. The winner was a coalition headed by the East German branch of West Germany's Christian Democratic Union, which advocated speedy reunification. After some negotiations (2+4 Talks) were held involving the two German states and the former Allied Powers which led to agreement on the conditions for German unification. The five original East German states that had been abolished in 1952 were recreated. On 3 October 1990, the five states officially joined the Federal Republic of Germany, while East and West Berlin united as a third city-state (in the same manner as Bremen and Hamburg). The great economic and socio-political inequalities between the former Germanies required government subsidy for the full integration of East Germany to the Federal German Republic.

Politics

SED logotype: The Communist–Social Democrat handshake of Wilhelm Pieck and Otto Grotewohl, establishing the SED in 1946.

DDR Flag At UN Headquarters, New York

There were four periods in East German political history.[22] These included: 1949–1961, which saw the building of socialism; 1961–1970 after the Berlin Wall closed off escape was a period of stability and consolidation; 1971–1985 was termed the Honecker Era, and saw closer ties with West Germany; and 1985–1989 saw the decline and extinction of East Germany.

Organization

The ruling political party in East Germany was the *Sozialistische Einheitspartei Deutschlands* (Socialist Unity Party of Germany, SED). It was created in 1946 through the Soviet-directed merger of the Communist Party of Germany (KPD) and the Social Democratic Party of Germany (SPD) in the Soviet controlled zone.

The Potsdam Agreement committed the Soviets to supporting a democratic form of government in Germany, and as in some other Warsaw Pact countries, other non-communist political parties were allowed. Nevertheless, every political party in the GDR had to register with the National Front of Democratic Germany, an alliance of parties and mass political organisations, including:

- *Christlich-Demokratische Union Deutschlands* (Christian Democratic Union of Germany, CDU), merged with the West-German CDU after reunification
- *Demokratische Bauernpartei Deutschlands* (Democratic Farmers' Party of Germany, DBD). This party was of special importance because of farmers' role in the economy. The party merged with the West German CDU after reunification.
- *Liberal-Demokratische Partei Deutschlands* (Liberal Democratic Party of Germany, LDPD), merged with the West German FDP after reunification.
- *Nationaldemokratische Partei Deutschlands* (National Democratic Party of Germany, NDPD), merged with the West German FDP after reunification.

Elections took place at the *Volkskammer* but were effectively controlled by the SED/state hierarchy, as Hans Modrow has noted. Elections were held in less-than-secret conditions, with voters given the choice of approving or rejecting "unity lists" put forward by the National Front and predetermining the distribution of seats given to the different parties and mass organisations. As was the case in most communist countries, approval rates of 90% or more were routine.

Voting in East Germany was relatively simple. To vote yes, a voter simply took the ballot paper, which contained only one name—that of the approved candidate—and dropped it into the voting box. A voter could vote against the candidate by crossing out his or her name, but had to do so in a separate voting booth without any secrecy. The consequences for such an act of defiance were severe—loss of one's job or expulsion from school, and close surveillance by the Stasi.[23]

The Volkskammer also included representatives from the *mass organisations* like the Free German Youth (*Freie Deutsche Jugend* or *FDJ*), or the Free German Trade Union Federation. In an attempt to include women in the political life of East Germany, there was a Democratic Women's Federation of Germany, with seats in the Volkskammer.

Palast der Republik, the seat of the Volkskammer

Important non-parliamentary mass organisations in East German society included the German Gymnastics and Sports Association (*Deutscher Turn- und Sportbund* or *DTSB*), and People's Solidarity (*Volkssolidarität*, an organisation for the elderly). Another society of note was the Society for German-Soviet Friendship.

Following German reunification, the SED was renamed the "Party of Democratic Socialism" (PDS) which subsequently merged with the West German WASG to form the Left Party (*Die Linke*). The Left Party continues to be a political force in many parts of Germany, albeit drastically less powerful than the SED.

Military

Because of East Germany's proximity to the West during the Soviet–American Cold War (1945–91), its four-branch military, the *Nationale Volksarmee* (National People's Army, NVA) was among the most advanced of the Warsaw Pact; its branches were:

1. Army *(Landstreitkräfte)*
2. Navy *(Volksmarine* – People's Navy)
3. Air Force–Air Defence *(Luftstreitkräfte/Luftverteidigung)*
4. GDR Border Troops *(Grenztruppen der DDR)*

The East German Military: A *Nationale Volksarmee* changing-of-the-guard ceremony, East Berlin.

Every man served eighteen months of compulsory military service; for the medically unqualified and the conscientious objector, there were the *Baueinheiten* construction units, established in 1964 in response to political pressure by the national Protestant Church upon the GDR's government. The armed forces of the GDR also possessed paramilitary reserve forces, such as the *Kampfgruppen der Arbeiterklasse* (Combat Groups of the Working Class), and from the Stasi, the Ministry for State Security (Mfs —*Ministerium für Staatssicherheit*), the "*Schild und Schwert der Partei*" (Shield and Sword of the Party).

Administrative districts

Until 1952, the GDR comprised the East Berlin capital city and the five German states of Mecklenburg-Vorpommern (in 1947 renamed: Mecklenburg), Brandenburg, Saxony-Anhalt, Thuringia, Saxony, their post-war territorial demarcations approximating the pre-war German demarcations of the Middle German *Länder* (states) and *Provinzen* (provinces of Prussia). Moreover, the western parts of two provinces, Pomerania and Lower Silesia, although prevailingly annexed by Poland, remained politically integral to the GDR.

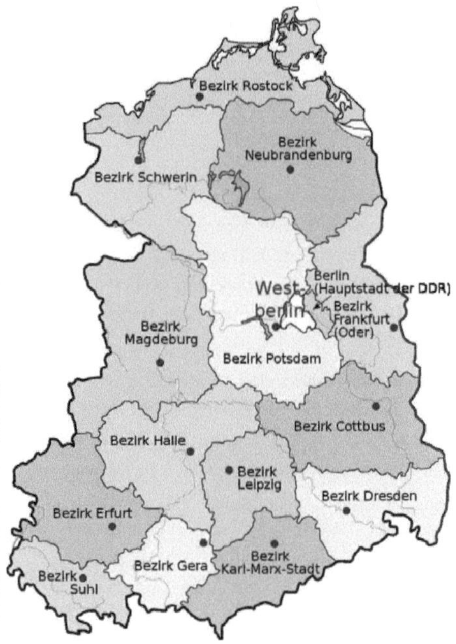

Administrative map: The districts of German Democratic Republic in 1952.

The East German Administrative Reform of 1952 disestablished the five states and established 14 *Bezirke* (districts), named per their capital cities: (i) Rostock, (ii) Neubrandenburg, (iii) Schwerin, (iv) Potsdam, (v) Frankfurt (Oder), (vi) Magdeburg, (vii) Cottbus, (viii) Halle, (ix) Leipzig, (x) Erfurt, (xi) Dresden, (xii) Karl-Marx-Stadt (Chemnitz until 1953 and again in 1990), (xiii) Gera, and (xiv) Suhl; East Berlin was denominated a *Bezirk* (district) in 1961.

The disestablished *Länder* (states), demarcated as *Bezirke* (districts), were renamed so: the northern *Land* (state) Mecklenburg was divided among the *Bezirke* Rostock, Schwerin, and Neubrandenburg; Brandenburg (containing Berlin) was divided into the Potsdam, Frankfurt (Oder), and Cottbus districts; Saxony-Anhalt was divided into the Halle and Magdeburg districts; the south-western state of Thuringia was divided into the Erfurt, Gera, and Suhl districts; and the south-eastern state of Saxony was divided among the Leipzig, Dresden, and Karl-Marx-Stadt districts.

East Berlin, the capital city of the German Democratic Republic, formed the *Bezirk* Berlin, the country's fifteenth district, retaining special legal status until 1968, when the residents voted in approving the new (draft) constitution. Regardless of the city's four-occupying-power-status, the ACC, and diplomatic objections of the Allied governments, the GDR administered the *Bezirk* Berlin as sovereign territory.

Major cities (1988 populations)

- Berlin, Hauptstadt der DDR (*English:* Berlin, Capital of the GDR) (1,200,000)
- Leipzig* (556,000)
- Dresden* (520,000)
- Karl-Marx-Stadt* (317,000) (Name now reverted to Chemnitz)
- Magdeburg* (290,000)
- Rostock* (250,000)
- Halle (Saale)* (236,000)
- Erfurt* (215,000)
- Potsdam* (140,000)
- Gera* (131,000)
- Schwerin* (130,000)
- Cottbus* (125,000)
- Zwickau (120,000)
- Jena (107,000)
- Dessau (105,000)

* *"Bezirksstadt" (centre of district)*

Uni-Riese (*University Giant*) in 1982
Built in 1972, it was once part of the Karl-Marx-University and is Leipzig's tallest building.

Population

The East German population declined by three million people throughout its forty-one year history, from 19 million in 1948 to 16 million in 1990; of the 1948 population, some 4 million were deported from the lands east of the Oder-Neisse line.[24] This was primarily a result of emigration — about one quarter of East Germans left the country before the Berlin Wall was completed in 1961, and after that time, East Germany had very low birth rates.[25] But in the years before reunification the birth rate in East Germany was much higher than in West Germany,[26] and in general the birth rate per woman was never much lower than in West Germany. This compares starkly with Poland, which increased during that time from 24 million in 1950 (a little more than East Germany) to 38 million (more than twice East Germany's population).

Economy

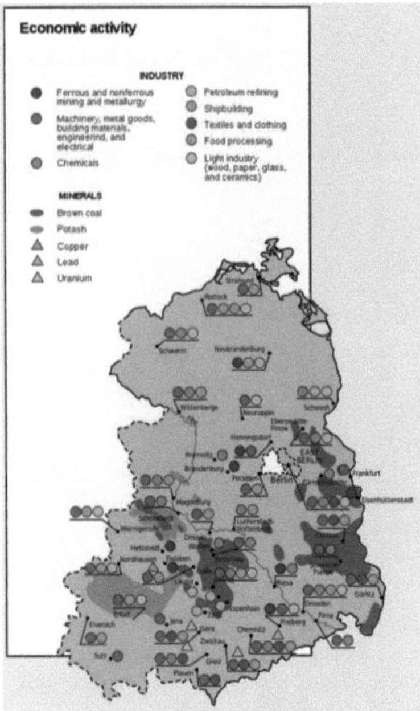

The East German economy

Communist economic staple: The Trabant automobile was a profitable product made in the German Democratic Republic (GDR).

The East German economy began poorly because of the devastation caused by the war, the loss of so many young soldiers, the disruption of business and transportation, the presence of so many refugees, and finally reparations owed to the USSR. The Red Army dismantled and transported to Russia the infrastructure and industrial plants of the Soviet Zone of Occupation. By the early 1950s, the reparations were paid in agricultural and industrial products; and Lower Silesia, with its coal mines and Szczecin, an important natural port, were given to Poland by the decision of Stalin.[16]

The socialist centrally planned economy of the German Democratic Republic was like that of the USSR. In 1950, the GDR joined the COMECON trade bloc. In 1985, collective (state) enterprises earned 96.7% of the net national income. To ensure stable prices for goods and services, the state paid 80% of basic supply costs. The estimated 1984 per capita income was $9,800 ($21,000 in 2008 dollars). In 1976, the average annual growth of the GDP was approximately five percent.[27]

Notable East German exports were photographic cameras, under the Praktica brand; automobiles under the Trabant, Wartburg, and the IFA brands; hunting rifles, sextants, and wristwatches.

Until the 1960s, East Germans endured shortages of basic foodstuffs such as sugar and coffee. East Germans with friends or relatives in the West (or any access to a hard currency) and the necessary Staatsbank foreign currency account, could afford Western products and export-quality East German products via Intershop. Consumer goods also were available, by post, from the Danish Jauerfood, and Genex companies.

The government used money and prices as political devices, providing highly subsidised prices for a wide range of basic goods and services, in what was known as "the second pay packet".[28] The economic results were highly negative, and created an increasing differential with the prosperity in West Germany. At the production level, artificial prices made for an inefficient, backward system of semi-barter and resource-hoarding. For the consumer, it led to the substitution of GDR money with time, barter, and hard-currencies. Ironically, the socialist economy became steadily more dependent on financial infusions from hard-currency loans from West Germany. East Germans, meanwhile, came to see their soft-currency as worthless relative to the Deutsche Mark (DM).[29]

Consumption and jobs

The existence of two German states created questions of legitimacy for both states, particularly in the 1950s when the outcome of the Cold War and the future of Germany remained much in doubt. Both German states used consumerism to promote their unique visions of *Germanness* (*Deutschtum*). For West Germany, the resumption of prewar patterns of consumption would signify a return to normality. This meant not only an end to the difficult shortages of the "hungry years," but also the curtailing of female employment outside the home.

In East Germany, people who were disgruntled resorted to nasty jokes at the expense of the oppressive regime; their names were recorded, and too many jokes would land a man in prison.[30]

Throughout its history the main criterion for getting a good job was unblinking loyalty to the Communist party bosses. Even in the electronics industry, a relatively modern and competitive sector of the GDR's economy, the criteria of professionalism were secondary to political criteria in personnel recruitment and development. Dubious loyalty meant exclusion from the university and from good jobs.

With a very low birth rate and a high rate of exodus, East Germany was losing workers. The solution was to import low-skilled workers from other Communist countries. Beginning in 1963 with a series of secret international agreements, East Germany recruited workers from Poland, Hungary, Cuba, Albania, Mozambique, Angola and (North) Vietnam. They numbered more than 100,000 by 1989. Their working conditions were bleak, and although they were officially equal to their German counterparts, the foreign workers remained at the bottom of the social ladder with almost no rights.[31]

Religion

Religion was contested ground in the GDR, with the Communists promoting atheism, although some people were loyal to Christian communities.[32]

Atheism

At the beginning, the Communist party had asserted the compatibility of Christianity and Marxism and sought Christian participation in the building of socialism. At first the question of atheism received little official attention. In the mid-1950s, as the Cold War heated up, atheism became suddenly a topic of major interest for the regime for propaganda purposes, both domestic and foreign. University chairs and departments devoted to the study of 'scientific atheism' were founded and much literature (scholarly and popular) on the subject was produced. Then this activity rather quickly subsided in the late 1960s amid perceptions that atheistic propaganda was becoming counterproductive; but official and scholarly attention to atheism was renewed beginning in 1973, though this time there was more emphasis on scholarship and the training of cadre than propaganda. Throughout, attention paid to atheism in East Germany always reflected politics and was never intended to jeopardise the cooperation that was desired from those East Germans who were religious.[33]

Protestants

East Germany historically was about 90% Protestant. Between 1956 and 1971 the leadership of the East German Lutheran Church changed its relations with the state from hostility to cooperation.[34] From the founding of the GDR in 1949, the Communist Party tried to weaken the influence of the church on the rising generation. The church therefore adopted an attitude of confrontation and distance regarding the Communist authorities. Around 1956 this firm stand against the regime began to wither in favour of a more neutral stance and conditional loyalty. The regime was no longer regarded as illegitimate; instead, the church leaders started viewing the authorities as installed by God and, therefore, deserving of obedience by Christians. But on matters where the state demanded something which was not in accordance with the will of God, the church reserved its right to say no. There were both structural and intentional causes behind this development. Structural causes included the hardening of Cold War tensions in Europe in the mid-1950s, which took away the temporary character of the East German state. The loss of church members and discrimination against young Christians also made it clear to the leaders of the church that they had to come into some kind of dialogue with the authorities. The intentions behind the change of attitude varied from a traditional Lutheran acceptance of secular power to a positive attitude toward socialist ideas. There was also a will to cooperate in order to have the ability to criticise from within a position of loyalty.[35]

Manfred Stolpe (b. 1936) became a lawyer for the Protestant church in 1959 before taking up a position at church headquarters in Berlin. In 1969 he helped found the Bund der Evangelischen Kirchen in der DDR, where he negotiated with the Communist government while at the same time working within the truly democratic system of the church's institutions. The international outlook he gained through the church's ecumenical activities helped him in his new job after winning the regional elections for the state of Brandenburg at the head of the SPD list in 1990. Despite accusations of having colluded with the Communist government, Stolpe, cleared of the charges, remained at the head of the Brandenburg government until he joined the federal government in 2002.

Catholics

The smaller Roman Catholic Church had a fully functioning episcopal hierarchy that was in full accord with the Vatican. During the early postwar years, tensions were high. The Catholic Church as a whole and particularly the bishops were resistant to both the regime and Marxist ideology, and the state allowed the bishops to lodge protests, which they did on issues such as abortion. The bishops were, however, closely observed by the Stasi.[35]

After 1945, the Church did fairly well in integrating Catholic exiles from lands to the east (which were given to Poland) and adjusting its institutional structures against the threats of an atheistic state. Within the Church, this meant an increasingly hierarchical structure, whereas in the area of religious education, press, and youth organisations, a system of temporary staff was developed, one that took into account the special situation of the Caritas, a charity organisation. They were hardly affected by Communist attempts to force them into line. By 1950, therefore, there existed a Catholic subsociety that was well adjusted to prevailing specific conditions and capable of maintaining Catholic identity.

With a generational change in the episcopacy taking place in the early 1980s, the state hoped for better relations with the new bishops, but the new bishops instead showed increasing independence from the state by holding unauthorised mass meetings, promoting international ties in discussions with theologians abroad, and hosting ecumenical conferences. The new bishops became less politically oriented and more involved in pastoral care and attention to spiritual concerns.

Culture

Music

The Puhdys and Karat were some of the most popular mainstream bands, managing to hint at critical thoughts in their lyrics without being explicit. Like most mainstream acts, they appeared in popular youth magazines such as *Neues Leben* and *Magazin*. Other popular rock bands were Wir, Dean Reed, City, Silly and Pankow. Most of these artists recorded on the state-owned AMIGA label.

Influences from the West were heard because West German TV and radio could be received in many parts of the East (an exception being Dresden with its geographical position in the Elbe valley, although limited reception of Western radio was still possible there). The Western influence led to the formation of more "underground" groups with a decisively western-oriented sound. A few of these bands were Die Skeptiker, Die Art and Feeling B. Additionally, hip hop culture reached the ears of the East German youth. With videos such as *Beat Street* and *Wild Style*, young East Germans were able to develop a hip hop culture of their own.[36] East Germans accepted hip hop as more than just a music form. The entire street culture surrounding rap entered the region and became an outlet for oppressed youth.[37]

Dissident singer/songwriter Wolf Biermann, although professing to be a dedicated Communist, was not permitted to perform in public, and was deprived of his East German citizenship in 1976, while visiting West Germany. He was not allowed to return to East Germany until 1989.

Singer Nina Hagen, who is the stepdaughter of Biermann, was exiled at the same time.

Governmental support of classical music maintained some fifty symphony orchestras, such as Gewandhausorchester and Thomanerchor in Leipzig; Sächsische Staatskapelle in Dresden; and Berliner Sinfonie Orchester and Staatsoper Unter den Linden in Berlin.

The birth place of Johann Sebastian Bach (1685–1750), Eisenach, was rendered as a museum about him, featuring more than three hundred instruments, which, in 1980, received some 70,000 visitors. In Leipzig, the Bach archive contains his compositions and correspondence and recordings of his music.

Theatre

East German theatre was originally dominated by Bertolt Brecht, who brought back many artists out of exile and reopened the *Theater am Schiffbauerdamm* with his Berliner Ensemble. Alternatively, other influences tried to establish a "Working Class Theatre", played for the working class by the working class.

Playwright Bertolt Brecht (1898–1956)

After Brecht's death, conflicts began to arise between his family (around Helene Weigel) and other artists about Brecht's heritage. Heinz Kahlau, Slatan Dudow, Erwin Geschonneck, Erwin Strittmatter, Peter Hacks, Benno Besson, Peter Palitzsch and Ekkehard Schall were considered to be among Bertolt Brecht's scholars and followers.

In the 1950s the Swiss director Benno Besson with the Deutsches Theater successfully toured Europe and Asia including Japan with *The Dragon* by Jewgenij Schwarz. In the 1960s, he became the Intendant of the Volksbühne often working with Heiner Müller.

After 1975 many artists left the GDR because of increasing censorship. A parallel theatre scene sprung up, creating theatre "outside of Berlin" in which artists played at provincial theatres. For example Peter Sodann founded the Neues Theater in Halle/Saale and Frank Castorf at the theater Anklam.

Theatre and cabaret had high status in the GDR, which allowed it to be very pro-active. This often brought it into confrontation with the state. Benno Besson once said, "In contrast to artists in the west, they took us seriously, we had a bearing."

Important theatres include: Berliner Ensemble;[38] – Deutsches Theater;[39] Maxim-Gorki-Theater;[40] and Volksbühne[41]

Volksbühne

Cinema

The prolific cinema of East Germany was headed by the DEFA,[42] *Deutsche Film AG*, which was subdivided in different local groups, for example *Gruppe Berlin*, *Gruppe Babelsberg* or *Gruppe Johannisthal*, where the local teams shot and produced films. Besides folksy movies, the movie-industry became known worldwide for its productions, especially children's movies (*Das kalte Herz*, film versions of the Brothers Grimm fairy tales and modern productions such as *Das Schulgespenst*).

Frank Beyer's *Jakob der Lügner* (Jacob the Liar), about persecution of Jews in the Third Reich, and *Fünf Patronenhülsen* (Five Cartridges), about resistance against fascism, became internationally famous.

Films about quotidian life, such as *Die Legende von Paul und Paula*, by Heiner Carow, and *Solo Sunny*, directed by Konrad Wolf and Wolfgang Kohlhaase, were very popular.

The film industry was remarkable for its production of *Ostern*, or Western-like movies. Indians in these films often took the role of displaced people who fight for their rights, in contrast to the American westerns of the time, where Indians were often either not mentioned at all or are portrayed as the villains. Yugoslavians were often cast as the Indians because of the small number of American Indians in eastern Europe. Gojko Mitić was well known in these roles, often playing the righteous, kindhearted and charming chief (*Die Söhne der großen Bärin* directed by Josef Mach). He became an honorary Sioux chief when he visited the United States in the 1990s, and the television crew accompanying him showed the tribe one of his movies. American actor and singer Dean Reed, an expatriate who lived in East Germany, also starred in several films. These films were part of the phenomenon of Europe producing alternative films about the colonization of America.

Because of censorship a certain number of very remarkable movies were forbidden at this time and reissued after the *Wende* in 1990. Examples are *Spur der Steine* (directed by Frank Beyer) and *Der geteilte Himmel* (directed by Konrad Wolf).

Cinemas in the GDR also showed foreign films. Czechoslovak and Polish productions were more common, but certain western movies were shown, though the numbers of these were limited because it cost foreign exchange to buy the licences. Further, movies representing or glorifying capitalistic ideology were not bought. Comedies enjoyed great popularity, such as the Danish *Olsen Gang* or movies with the French comedian Louis de Funès.

Since the fall of the Berlin Wall, several movies depicting life in the GDR have been critically acclaimed. Some of the most notable were *Good Bye Lenin!* by Wolfgang Becker, *Das Leben der Anderen* (The Lives of Others) by Florian Henckel von Donnersmarck (won the Academy Award for best Film in a Foreign Language) in 2006,[43] *Alles auf Zucker!* (Go for Zucker) by Dani Levi. Each film is heavily infused with cultural nuances unique to life in the GDR.

Sport

East Germany was very successful in the sports of bicycling, weight-lifting, swimming, gymnastics, track and field, boxing, ice skating, and winter sports. The success is attributed to the leadership of Dr. Manfred Hoeppner which started in the late 1960s.

Another supporting reason was doping in East Germany, especially with anabolic steroids, the most detected doping substances in IOC-accredited laboratories for many years.[44] [45] The development and implementation of a state-supported sports doping program helped East Germany, with its small population, to become a world leader in sport during the 1970s and 1980s, winning a large number of Olympic and world gold medals and records.[46]

Another factor for success was the furtherance-system for young people in GDR. Sport-teachers at school were encouraged to look for certain talents in children ages 6 to 10 years old. For older pupils it was possible to attend grammar-schools with a focus on sports (for example sailing, football and swimming). This policy was also used for talented pupils with regard to music or mathematics.

Sports clubs were highly subsidized, especially sports in which it was possible to get international fame. For example, the major leagues for ice hockey and basketball just included each 2 teams (excluding the school and university sport), though a lack of popular interest in both ice hockey and basketball was and is common across the whole of Europe, whether communist or not. Football (soccer) was the most popular sport. Club football teams such as Dynamo Dresden, 1. FC Magdeburg, FC Carl Zeiss Jena, 1. FC Lokomotive Leipzig and BFC Dynamo had successes in European competition. Many East German players such as Matthias Sammer and Ulf Kirsten became integral parts of the reunified national football team. Other sports enjoyed great popularity like figure skating, especially because of sportspeople like Katarina Witt.

Karin Janz

Successful athletes

- Waldemar Cierpinski, athlete
- Ernst Degner, racing motorcyclist
- Heike Drechsler, athlete
- Maxi Gnauck, gymnast
- Lutz Heßlich, track cyclist
- Falk Hoffmann, diver
- Jan Hoffmann, ice skater
- Karin Janz, gymnast
- Karin Kania, speed skater
- Marita Koch, athlete
- Christa Luding-Rothenburger, speed skater and track cyclist
- Olaf Ludwig, road cyclist
- Henry Maske, boxer
- Heinz Melkus, auto racing driver
- Peter Muecke, rally driver
- Meinhard Nehmer, bobsledder
- Frank-Peter Roetsch, biathlete
- Gustav-Adolf Schur, road cyclist
- Gaby Seyfert, ice skater
- Jürgen Sparwasser, footballer
- Jens Weißflog, ski jumper
- Fred Williamowski, rallying motorcyclist

Track-suit diplomacy

The East and the West also competed via sport; GDR athletes dominated several Olympic sports. Of special interest was the only football match between the Federal Republic of Germany and the German Democratic Republic, a first-round match during the 1974 FIFA World Cup, which the East won 1–0, although West Germany, the host, was the eventual champion.

Television and radio

Television and radio in East Germany were state controlled; the *Rundfunk der DDR* was the official radio broadcasting organisation from 1952 until German reunification. The organization was based in the *Funkhaus Nalepastraße* in East Berlin. *Deutscher Fernsehfunk* (DFF), from 1972–1990 known as *Fernsehen der DDR* or DDR-FS, was the state television broadcaster from 1952. Reception of Western radio (and even television) broadcasts was widespread.[47]

"25 years of the GDR" is a 1974 postage stamp commemorating the 25th anniversary of East Germany's establishment on 7 October 1949.

Telecommunications

Further information: Communications in Germany

By the mid-1980s, East Germany possessed a well-developed communications system. There were approximately 3.6 million telephones in usage (21.8 for every 100 inhabitants), and 16,476 Telex stations. Both of these networks were run by the Deutsche Post der DDR (East German Post Office). East Germany was assigned telephone country code 37; in 1991, several months after reunification, East German telephone exchanges were incorporated into country code 49.

1989 USSR stamp: "40 years of the German Democratic Republic"

An unusual feature of the telephone network was that in most cases, direct dialing for long distance calls was not possible. Although area codes were assigned to all major towns and cities, they were only used for switching international calls. Instead, each location had its own list of dialing codes with shorter codes for local calls and longer codes for long distance calls. After reunification, the existing network was largely replaced, and area codes and dialing became standardised.

In 1976 East Germany inaugurated the operation of a ground-based radio station at Fürstenwalde for the purpose of relaying and receiving communications from Soviet satellites and to serve as a participant in the international telecommunications organization established by the Soviet government, Intersputnik.

Holidays

Date	English Name	German Name	Remarks
1 January	New Year's Day	Neujahr	
Moveable feast	Good Friday	Karfreitag	
Moveable feast	Easter Sunday	Ostersonntag	
Moveable feast	Easter Monday	Ostermontag	Was not an official holiday after 1967.
1 May	May Day	Tag der Arbeit	International Workers' Day
8 May	Victory in Europe Day	Tag der Befreiung	The translation means "Day of Liberation"
Moveable feast	Father's Day / Ascension Day	Vatertag / Christi Himmelfahrt	Thursday after the 5th Sunday after Easter. Was not an official holiday after 1967.

Moveable feast	Whitmonday	Pfingstmontag	50 days after Easter Sunday
7 October	Republic Day	Tag der Republik	National holiday
Moveable feast	Day of Repentance and Prayer	Buß- und Bettag	Penultimate Wednesday before the fourth Sunday before December 25. Was not an official holiday after 1967.
25 December	First Day of Christmas	1. Weihnachtsfeiertag	
26 December	Second Day of Christmas	2. Weihnachtsfeiertag	

Ostalgie

The end of the Cold War division of Germany and unification in 1990 inspired initial euphoria.[48]

But for many East Germans, this joy quickly turned to dismay. West Germans often acted as if they had "won" and East Germans had "lost" in unification, leading many East Germans (*Ossis*) to resent West Germans (*Wessis*). Additionally, the dislocations associated with the end of Communism, the disappearance of East Germany and German unification were hardest for East Germany, where unemployment skyrocketed and many East German professionals quickly fled for better jobs in West Germany.

These and other effects of unification led many East Germans to begin to think of themselves more strongly as "East" Germans rather than as simply as "Germans". This produced in many former GDR citizens a longing for certain aspects of the former East Germany, such as full employment and other perceived benefits of the GDR state, termed "Ostalgie" (a blend of *Ost* "east" and *Nostalgie* "nostalgia") and depicted in the Wolfgang Becker film *Goodbye Lenin!*.

Danish historian Feiwel Kupferberg (2002) argues that the real difficulty in German reunification was the discrepancy in the ways the West Germans and East Germans have viewed their Nazi past. The West Germans grimly faced it, atoned for it, and transformed their half of the country into a prosperous, free democracy that valued both individual freedom and responsibility. By contrast, the East Germans absorbed the Soviet-made myth that East Germany was the "victor of history" that successfully resisted the fascists. They blamed their western compatriots for the Nazi atrocities because West Germany had a capitalist economy, which had created Hitler; without capitalism, East Germany was now pure. This attitude, Kupferberg argues, allowed middle classes that had supported Hitler to retain powerful roles in East Germany and there was no radical break with the Nazi regime.[49]

See also

Germany

- Berlin
  - Berlin Wall
  - East Berlin
  - West Berlin
- History of East Germany
- History of Germany since 1945
- Inner German border
- Iron Curtain
- Leaders of East Germany
- Ministerrat
- West Germany

Armed Forces

- Conscientious objection in East Germany
- Grenztruppen
- Landstreitkräfte
- Luftstreitkräfte
- National People's Army
- Stasi
- Volksmarine
- Volkspolizei

Media

- *Aktuelle Kamera*, GDR's main TV news show
- Broadcasting in East Germany
- *Der Tunnel*, a film about a mass evacuation to West Berlin through a tunnel
- *Good Bye, Lenin!*, a tragicomedy film about the German reunification
- Radio Berlin International

Transport

- Barkas
- Deutsche Reichsbahn – The railway company of the GDR
- Interflug – The airline of the GDR
- Trabant
- Transport in the German Democratic Republic
- Wartburg

Other

- Education in the German Democratic Republic
- Fichtelberg, highest point (1214.6m)
- Index of East Germany-related articles
- GDR jokes
- Ostalgie (Nostalgia, missing the old days.)
- Palast der Republik
- Dean Reed
- Sportvereinigung (SV) Dynamo
- Tourism in East Germany
- Omoiyari Yosan (DDR government→USSR Forces

Notes

[1] Top-Level-Domain .DD (http://internet.robert-scheck.de/tld-dd/) Information site about .dd in German language

[2] The last division: a history of Berlin, 1945-1989, Ann Tusa, p.73

[3] However, the usage of the abbreviation *BRD* (FRG) for the West German, *Bundesrepublik Deutschland* (*Federal Republic of Germany*), was never accepted in West Germany since it was considered a political statement. Thus BRD (FRG) was a term used by East Germans, or it might have been used by West Germans who held a pro-East German view. Colloquially, West Germans called West Germany simply Germany, or, alternatively, *Bundesrepublik* or *Bundesgebiet* (federal republic, or federal territory, respectively), referring to the country and *Bundesbürger* (federal citizen[s]) for its citizens, with the adjective, *bundesdeutsch* (federally German).

[4] Gerhard A. Ritter, "Die DDR in der Deutschen Geschichte," [The GDR in German history] *Vierteljahrshefte für Zeitgeschichte,* Apr 2002, Vol. 50 Issue 2, pp 171–200; this author is the son of historian Gerhard Ritter.

[5] "Yalta Conference" (http://www.spartacus.schoolnet.co.uk/2WWyalta.htm). spartacus.schoolnet.co.uk. . Retrieved 25 September 2010.

[6] See Spilker (2006)

[7] See Anna M. Cienciala "History 557 Lecture Notes (http://web.ku.edu/~eceurope/hist557/lect17.htm)

[8] David S. Wyman and Charles H. Rosenzveig, *The world reacts to the Holocaust* (1996) p 457-66

[9] "State symbols: the quest for legitimacy in the Federal Republic of Germany and the German Democratic Republic, 1949-1959", by Margarete Myers Feinstein, page 78: " (...) claims of East Berlin as the capital of the GDR (...), East Berlin was not recognized by the West and most Third World countries"

[10] Michael D. Haydock, *City Under Siege: The Berlin Blockade and Airlift, 1948–1949* (2000)

[11] Weitz 1997, p. 350: Following a Soviet order in February 1948, the German Economic Commission became a nascent state structure for all intents and purposes, with competence far beyond the economy proper and it was granted power to issue orders and directives to all German organs within the Soviet Occupation Zone.

[12] McCauley 1983, p. 38: It (the DWK) had become the de facto government of the Soviet zone. Its chairman was Heinrich Rau (SED) and four of his six deputies were also SED members.

[13] Patrick Major and Jonathan Osmund, *Workers' and Peasants' State: Communism and Society in East Germany under Ulbricht, 1945–71* (2002)

[14] East Berlin 17 June 1953: Stones Against Tanks (http://www.dw-world.de/dw/article/0,,894998,00.html), *Deutsche Welle,* Accessed 16-May-2007

[15] Victor Baras, "Beria's Fall and Ulbricht's Survival," *Soviet Studies,* 1975, Vol. 27 Issue 3, pp 381–395

[16] Norman M. Naimark. *The Russians in Germany: A History of the Soviet Zone of Occupation, 1945–1949.* Harvard University Press, 1995. ISBN 0-674-78405-7 pp. 167–9

[17] Frederick Taylor, *Berlin Wall: A World Divided, 1961–1989* (2007)

[18] Henry Krisch, "Soviet-GDR Relations in the Honecker Era," *East Central Europe,* Dec 1979, Vol. 6 Issue 2, pp 152–172

[19] EAST GERMANY: The Price of Recognition, Time Magazine, Jan. 01, 1973 (http://www.time.com/time/magazine/article/0,9171,903634,00.html)

[20] The Berlin Wall (1961–89) (http://www.germannotes.com/hist_east_wall.shtml) German Notes. Retrieved 2006-10-24.

[21] Darnton, Robert, *Berlin Journal* (New York, 1992, W.W. Norton) pp.98–99

[22] David P. Conradt, *The German Polity* (2008) p. 20

[23] Sebetsyen, Victor (2009). *Revolution 1989: The Fall of the Soviet Empire*. New York City: Pantheon Books. ISBN 0375425322.

[24] "East Germany: country population" (http://www.populstat.info/Europe/germanec.htm). Populstat.info. . Retrieved 2010-03-28.

[25] "Germany Population – Historical Background" (http://www.country-studies.com/germany/population---historical-background.html). Country-studies.com. . Retrieved 2010-03-28.

[26] Destatis.de (http://www.destatis.de/jetspeed/portal/cms/Sites/destatis/Internet/DE/Content/Publikationen/Fachveroeffentlichungen/Bevoelkerung/BroschuereGeburtenDeutschland,property=file.pdf) page 17

[27] "Business America. (27 February 1989). German Democratic Republic: long history of sustained economic growth continues; 1989 may be an advantageous year to consider this market – Business Outlook Abroad: Current Reports from the Foreign Service." (http://findarticles.com/p/articles/mi_m1052/is_n4_v110/ai_7373809). *Business America*. 1989. . Retrieved 2007-10-02.

[28] Boroch, Wilfried (1996), "Social policy as an institutional transformation problem", *Transition Economies*, Volume 31, Number 3, pp139-146

[29] Jonathan R. Zatlin, *The Currency of Socialism: Money and Political Culture in East Germany* (2007)

[30] Jonathan R. Zatlin, *The Currency of Socialism: Money and Political Culture in East Germany*. (2007).

[31] Raymond G. Stokes, *Constructing Socialism: Technology and Change in East Germany, 1945–1990* (2000)

[32] Mary Fulbrook, "The Limits Of Totalitarianism: God, State and Society in the GDR," *Transactions of the Royal Historical Society*, Jan 1957, Vol. 7 Issue 1, pp 25–52

[33] Fulbrook, "The Limits Of Totalitarianism: God, State and Society in the GDR"

[34] Martin Onnasch, "Konflikt Und Kompromiss: Die Haltung Der Evangelischen Kirchen Zu Den Gesellschaftlichen Veränderungen In Der Ddr Am Anfang Der Fünfziger Jahre," ["Conflict and compromise: the position of the Protestant churches with regard to social changes in the GDR at the beginning of the 1950s"], *Kirchliche Zeitgeschichte / Halbjahresschrift fuer Theologie und Geschichtswisseschaft*, 1990, Vol. 3 Issue 1, pp 152–165

[35] Stephen R. Bowers, "Private Institutions in Service to the State: The German Democratic Republic'S Church in Socialism," *East European Quarterly*, Spring 1982, Vol. 16 Issue 1, pp 73–86

[36] Brown, Timothy S. "'Keeping it Real' in a Different 'Hood: (African-) Americanization and Hip-hop in Germany." In The Vinyl Ain't Final: Hip Hop and the Globalization of Black Popular Culture, ed. by Dipannita Basu and Sidney J. Lemelle, pp.137–150. London; A

[37] Elfein, Dietmar. From Krauts with Atittudes to Turks with Attitudes: Some a Aspects of Hip-Hop History in Germany. pp.225–265 Popular Music vol. 17:3. October 1998.

[38] "Das BE – ein Theater für Zeitgenossen" (http://www.berliner-ensemble.de). Berliner-ensemble.de. . Retrieved 2010-03-28.

[39] "Deutsches Theater: Home" (http://www.deutsches-theater.de). Deutsches-theater.de. . Retrieved 2010-03-28.

[40] "Gorki.de" (http://www.gorki.de). Gorki.de. . Retrieved 2010-03-28.

[41] "Volksbühne Berlin" (http://www.volksbuehne-berlin.de). Volksbuehne-berlin.de. . Retrieved 2010-03-28.

[42] "DEFA – Stiftung – Home" (http://www.defa-stiftung.de). Defa-stiftung.de. . Retrieved 2010-03-28.

[43] 2006 Academy Award for "The Lives of Others" (http://www.imdb.com/title/tt0405094/awards),

[44] Hartgens and Kuipers (2004), p. 515

[45] Kicman AT, Gower DB (July 2003). "Anabolic steroids in sport: biochemical, clinical and analytical perspectives" (http://acb.rsmjournals.com/cgi/pmidlookup?view=long&pmid=12880534). *Annals of clinical biochemistry* **40** (Pt 4): 321–56. doi:10.1258/000456303766476977. PMID 12880534. .

[46] Tagliabue, John. – "Political Pressure Dismantles East German Sports Machine" (http://query.nytimes.com/gst/fullpage.html?res=9D0CE7D81039F931A25751C0A967958260&sec=&spon=&pagewanted=2) – *New York Times* – 12 February 1991 | Janofsky, Michael. – "OLYMPICS; Coaches Concede That Steroids Fueled East Germany's Success in Swimming" (http://query.nytimes.com/gst/fullpage.html?sec=health&res=9D0CE1DA1731F930A35751C1A967958260) – *New York Times* – 3 December 1991 | Kirschbaum, Erik. – "East German dope still leaves tracks" (http://www.rediff.com/sports/2000/sep/15germ.htm) – Rediff from Reuters – 15 September 2000 | Ungerleider, Steven (2001). *Faust's Gold: Inside The East German Doping Machine*. Thomas Dunne Books ISBN 0-312-26977-3 | "Little blue pills and a lot of gold..." (http://www.shorel.com/faust.cfm) - Shorel.com | Culture & Lifestyle: "Sports Doping Statistics Reach Plateau in Germany" (http://www.dw-world.de/dw/article/0,2144,786574,00.html) – *Deutsche Welle* – 26 February 2003 | "The East German Doping Machine" (http://www.ishof.org/honorees/german_doping.htm) – International Swimming Hall of Fame | Culture & Lifestyle: "East Germany's Doping Legacy Returns" (http://www.dw-world.de/dw/article/0,2144,1344595,00.html) – *Deutsche Welle* – 10 January 2004 | Longman, Jere. – "East German Steroids' Toll: 'They Killed Heidi'" (http://www.nytimes.com/2004/01/26/sports/othersports/26STER.html?ei=5007&en=bc039fed4c6ea23a&ex=1390453200&partner=USERLAND&pagewanted=print&position=) – *New York Times* – 26 January 2004 | Harding, Luke. – "Forgotten victims of East German doping take their battle to court" (http://www.guardian.co.uk/sport/2005/nov/01/athletics.gdnsport3) – *The Guardian* – 1 November 2005 | Jackson, Guy. Winning at Any Cost?: "Doping for glory in East Germany" (http://portal.unesco.org/en/ev.php-URL_ID=33956&URL_DO=DO_TOPIC&URL_SECTION=201.html) – UNESCO – September 2006 | "Ex-East German athletes compensated for doping" (http://sports.espn.go.com/oly/news/story?id=2696038) – Associated Press – (c/o ESPN) – 13 December 2006 | "East German doping victims to get compensation" (http://www.cbc.ca/sports/amateur/story/2006/12/13/geramny-doping.html) – Associated Press – (c/o CBC Canadian Broadcasting Corporation) – 13 December 2006 | Starcevic, Nesha. – "East German doping victims to get compensation" (http://www.signonsandiego.com/sports/20061213-0605-germanydoping-compensation.html) – Associated Press – (c/o *San Diego Union-Tribune*) – 13 December 2006 | "Germany completes $4.1M payout to doping victims" (http://www.usatoday.com/sports/olympics/

2007-10-11-doping-damages_N.htm?csp=34) – *USA Today* – 11 October 2007 | "East Germany's Secret Doping Program" (http://www.thirteen.org/scienceandnature/east-germanys-secret-doping-program) – *Secrets of the Dead* – Thirteen/WNET – 7 May 2008

[47] *Representing East Germany since unification: from colonization to nostalgia* (http://books.google.com/books?id=H7ajL_aNu6oC&pg=PA146), By Paul Cooke, Berg Publishers, 1 August 2005, ISBN 978-1-84520-189-0, page 146. Retrieved from Google Books Jan 25, 2010.

[48] Martin Blum, "Remaking the East German Past: 'Ostalgie,' Identity, and Material Culture," *Journal of Popular Culture*, Winter 2000, Vol. 34 Issue 3, pp 229–54

[49] Feiwel Kupferberg, *The Rise and Fall of the German Democratic Republic* (2002), pp 54–55, 66, 94–5, 192–3.

References and bibliography

- Allinson, Mark. *Politics and Popular Opinion in East Germany 1945–68* (2000)
- Augustine, Dolores. *Red Prometheus: Engineering and Dictatorship in East Germany, 1945–1990.* (2007) 411pp
- Baylis, Thomas A., David H Childs and Marilyn Rueschemeyer, eds.; East Germany in Comparative Perspective, Routledge. 1989
- Childs, David H.. *The Fall of the GDR,* Longman Personed.co.uk (http://www.pearsoned.co.uk/Bookshop/detail.asp?item=100000000004463), 2001. ISBN13:9780582315693, ISBN10: 0582315697
- Childs, David H.. & Richard Popplewell. *The Stasi: East German Intelligence and Security Service,* Palgrave Macmillan Palgrave.com (http://www.palgrave.com/products/title.aspx?is=0333772075), Amazon.co.uk (http://www.amazon.co.uk/Stasi-Intelligence-Security-Service-1917-89/dp/0333772075) 1996.
- Childs, David H.. *The GDR: Moscow's German Ally,* George Allen & Unwin, 1983. ISBN 0-04-354029-5, 9780043540299.
- Childs, David H.. *The Two Red Flags: European Social Democracy & Soviet Communism Since 1945,* Routledge, 2000. Informaworld.com (http://www.informaworld.com/smpp/title~db=all~content=t732627883)
- Fulbrook, Mary. The People's State: East German Society from Hitler to Honecker Yale University Press, 2005. 352 pp. ISBN 0-300-10884-2.
- Fulbrook; Mary. Anatomy of a Dictatorship: Inside the GDR, 1949–1989 Oxford University Press, 1995
- Gray, William Glenn. *Germany's Cold War: The Global Campaign to Isolate East Germany, 1949–1969* University of North Carolina Press. 2003
- Grix, Jonathan. *The Role of the Masses in the Collapse of the GDR* Macmillan, 2000
- Jarausch, Konrad H., and Eve Duffy; *Dictatorship as Experience: Towards a Socio-Cultural History of the GDR* Berghahn Books, 1999
- Kupferberg, Feiwel. *The Rise and Fall of the German Democratic Republic* (2002) 228pp; online review (http://www.h-net.org/reviews/showrev.php?id=7461)
- McCauley, Martin (1983). *The German Democratic Republic since 1945* (http://www.google.com/search?hl=de&tbo=1&tbs=bks:1&q="Heinrich+Rau"+DWK+"de+facto+government"&aq=f&aqi=&aql=&oq=&gs_rfai=). Palgrave Macmillan. ISBN 0-333-26219-0. Retrieved 2010-10-24.
- Naimark, Norman M. *The Russians in Germany: A History of the Soviet Zone of Occupation, 1945–1949* (1997) excerpt and text search (http://www.amazon.com/Russians-Germany-History-Occupation-1945-1949/dp/0674784065/)
- Port, Andrew I. *Conflict and Stability in the German Democratic Republic* Cambridge University Press, 2007.
- Pritchard, Gareth, *The Making of the GDR 1945–53: From Antifascism to Stalinism* (2000)
- Ritter, Gerhard A. "Die DDR in der Deutschen Geschichte," [The GDR in German history] *Vierteljahrshefte für Zeitgeschichte,* Apr 2002, Vol. 50 Issue 2, pp 171–200
- Steiner, André. *The Plans That Failed: An Economic History of East Germany, 1945–1989* (2010)
- Spilker, Dirk. *The East German Leadership and the Division of Germany: Patriotism and Propaganda 1945–1953.* (2006). online review (http://www.h-net.org/reviews/showrev.php?id=23832)
- Stokes, Raymond G. *Constructing Socialism: Technology and Change in East Germany, 1945–1990* (2000)

- Weitz, Eric D. (1997). *Creating German communism, 1890–1990* (http://books.google.com/?id=5kkQyWIFCbIC&pg=PA350&lpg=PA350&dq="German+Economic+Commission"+"Deutsche+Wirtschaftskommission"&q="German Economic Commission" "Deutsche Wirtschaftskommission"). Princeton University Press. ISBN 0-691-02594-0.
- Zatlin, Jonathan R. *The Currency of Socialism: Money and Political Culture in East Germany.* (2007). 377 pp. online review (http://www.h-net.org/reviews/showrev.php?id=22672)

External links

- AHF – Nationale Volksarmee (NVA) (http://www.axishistory.com/index.php?id=5528)
- (German) *Auferstanden aus Ruinen* (http://www.auferstanden-aus-ruinen.de/)
- (German) *That was the GDR (in German)* (http://www.ddr.innerdeutsche-grenze.info/)
- Translations of propaganda materials from the GDR. (http://www.calvin.edu/cas/gpa/gdrmain.htm)
- (German) Geschichte des ostdeutschen Designs – history of east German design (http://www.industrieform-ddr.de/)
- DDR Museum Berlin – Culture of the GDR (http://www.ddr-museum.de/en/)
- East Berlin, Past and Present (http://www.ostberlin.de/en/)
- Pictures of the GDR 1949–1973 (http://www.ddr-fotos.de/gdr_photos.htm)
- RFE/RL East German Subject Files (http://www.osaarchivum.org/db/fa/300-3-1.htm) Open Society Archives, Budapest
- Stamps Catalog of the German Democratic Republic (http://colnect.com/en/stamps/series/country/2979)
- East German anthem with English and German lyrics (http://eastgermany.info/EastGermanAnthem.htm)

Countries of the world · Europe

kbd:Демократ Джэрмэн Республикэ

Treptower Park

Treptower Park (pronounced [tʁɛptoːɐ], with a silent *w*) is a park along the river Spree in Alt-Treptow, in the district of Treptow-Köpenick, south of central Berlin. The park is a popular place for recreation of Berliners and a tourist attraction. It was also used for the concert given by the British Rock Band Barclay James Harvest on the 14th July 1987, which was the first ever open-air concert by a Western Rock Band in the German Democratic Republic[1]

Soviet war memorial

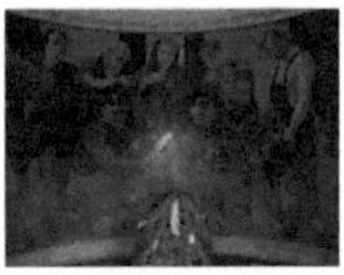

The statue of the Soviet soldier holding a child and stamping on a crushed Swastika (left) and a view into the pedestal

Its prominent feature is the Soviet War Memorial (sometimes translated as the "Soviet Cenotaph"), built to the design of the Soviet architect Yakov Belopolsky to commemorate the 50,000 Soviet soldiers who fell in the Battle in Berlin in April-May 1945. It was opened four years after the war ended on May 8, 1949.

Spreepark

Within Treptower Park, there can be found an abandoned amusement park, known as 'Spree Park - Berlin'. The amusement park operated from October 1969 to 2001.

The owner of Spree Park, Norbert Witte went bankrupt and left Germany rather abruptly. He took several of the park's rides including the Jet Star and Fun Express with him to Peru where he was to open a small park called Lunapark at Jockey Plaza in Lima, Peru. Some of the rides were damaged in transit and then followed by legal discussions of who would pay for the damages.

References

[1] Barclay James Harvest Biography (http://www.bjharvest.co.uk/bjh-biog.htm)

External links

- Website der Bürgerinitiative Treptower Park (http://www.treptowerpark.de/)
- Rollercoaster Database - Spree-Park Info (http://www.rcdb.com/pd347.htm)

Berlin Warschauer Straße station

DB	
Warschauer Straße	

S-Bahnhof Warschauer Straße

Operations	
Category	4
Type	Bf
DS100 code	BWRS
Station code	6550
Construction and location	
Location	Berlin
State	Berlin
Country	Germany
Local authority	Friedrichshain-Kreuzberg
List of railway stations in the Berlin area	

Berlin Warschauer Straße station is an S-Bahn and U-Bahn station located on Warschauer Straße in the Friedrichshain locality of central Berlin, Germany.

Overview

U-Bahnhof Warschauer Straße

The U-Bahn station is the eastern terminus of the Berlin U-Bahn line U1. Opened in 1902 under the name *Warschauer Brücke* as one of the first stations on the network, it is an above-ground station. It is the only station of the U1 located in the eastern part of the city and was closed in 1961 following the construction of the Berlin Wall. The adjacent Stralauer Tor station was also located in the borough of Friedrichshain, but it was destroyed in 1945 and not rebuilt. The U-Bahn station was reopened on October 14, 1995 following extensive reconstruction. At the same time it was renamed *Warschauer Straße* in order to create uniformity with the adjacent S-Bahn station.

The central S-Bahn platform, looking west

In the basement vaults of the building three discotheqes are located: The Matrix Club, since 1996, one of the biggest venues in Berlin with up to nine bars and five dancefloors, the *Narva Lounge* and the *Busche*.

Street sign of Warschauer Straße in Berlin

The Berlin S-Bahn station of the same name is around 150 metres away. Unique is the third platform, added in 1986, allowing the passengers to get off the train on the left and right. The S-Bahn station has been undergoing reconstruction since 2004, the entrance building was demolished in 2005.

Both stations are known collectively as *S+U Warschauer Straße*. Plans call for the existing U-Bahn station to be closed and the line to be extended over a bridge to reach the S-Bahn station, to create a cruciform transfer station. However, due to the precarious financial situation in Berlin it is uncertain whether this part of the development will take place. A further 400-m extension of the U1 to intersect with the U5 at Frankfurter Tor has been planned since the U5 was being constructed in the 1920s, but is unlikely to be built in the near future.

Warschauer Straße station is located on the eponymous street named after Warsaw, on the northeastern bank of the river Spree. The Oberbaumbrücke, the East Side Gallery as well as the new O2 World arena can be reached on foot.

Preceding station	**Berlin S-Bahn**	**Following station**
Ostbahnhof *toward Spandau*	S3	Ostkreuz *toward Erkner*
Ostbahnhof *toward Westkreuz*	S5	Ostkreuz *toward Strausberg Nord*
Ostbahnhof *toward Potsdam Hbf*	S7	Ostkreuz *toward Ahrensfelde*
Ostbahnhof *toward Spandau*	S75	Ostkreuz *toward Wartenberg*
Preceding station	**Berlin U-Bahn**	**Following station**
Schlesisches Tor *toward Uhlandstraße*	U1	*Terminus*

Article Sources and Contributors

Ostkreuz *Source*: http://en.wikipedia.org/w/index.php?title=Ostkreuz *Contributors*: AalanTG, Anthony Appleyard, Boneka, DerBorg, Dinamik, Doco, IsarSteve, Jumbo1435, Kusma, Maury Markowitz, Müslimix, Nicolaiplum, PKT, ProhibitOnions, Starbois, Valenciano, Xyboi, 3 anonymous edits

Berlin *Source*: http://en.wikipedia.org/w/index.php?title=Berlin *Contributors*: (jarbarf), 119, 1988ja, 1brettsnyder, 2112 rush, 334a, 52 Pickup, A.h. king, A3RO, ABF, AEMoreira042281, APPER, Aaadreri, Aara5250123, Aaron Einstein, Abani79, Acalamari, Acather96, Acela Express, Aceman9188, Acntx, Adam Carr, Adambiswanger1, AdjustShift, Afhaalchinees, Ahm922, Ahoerstemeier, Ahuskay, Aksi great, Alagemo, AlanFord, Alandeus, Alansohn, Albamo, Ale jrb, AlexCruise, AlexiusHoratius, Alexvgtb, Ali@gwc.org.uk, Alphasinus, Altenmann, Alterego, Amazonien, Amplitude101, Ams80, Anarchovegan, Ancheta Wis, Andie Gilmour, Andre Engels, AndreHuppertz, Andrea.gf, Andrew0921, AndrewHowse, Andrewlp1991, Andrewpmk, Andromeda, Andrwsc, Andy Marchbanks, Andycjp, Angela, Angr, Anon user, Anonymous Dissident, Anorak2, Antandrus, Anthony Appleyard, Antitext, Antman100, Anville, Argos'Dad, Aribfes, ArkansasTraveler, Arminia, ArtVandelay13, Artystarty, Astral, Athinaios, Atif.t2, Atlant, Attilios, Aufbauten, Avala, Averell23, Averette, Avicennasis, AxG, Axt, BLueFiSH.as, BRG, Bact, Baldhur, Barliner, Baronnet, Barry567, Bart133, Baxterguy, Bayerischermann, Bcnviajero, Bearcat, Beenjuity, Beetstra, Belamp, Bellwars, Bentley4, Benw1212, Berek, Berlin 11011, Berlininlondon, Bert166, Bhoesicol, Big Moira, Bigbossfarin, Biglatrell, BishkekRocks, Bjarki S, Blartyfarty, Blathnaid, Blaxthos, Bletch, Blueplane, Bluerasberry, Blur4760, Bob Burkhardt, Bobblehead, Bobblewik, Bobmoney711, Bobo192, Bobrayner, Bojin, Bongwarrior, Boothy443, Boson, Bowl2021, Boysporting, Brandmeister (old), Brandon1978, Brian0918, Brianjd, Brion VIBBER, BsBsBs, Bsadowski1, Bucketsofg, Burningview, Burschenschafter, Bushy moustache, CORNELIUSSEON, Caknuck, CalJW, CalendarWatcher, Calréfa Wénä, Caltas, Cameron Nedland, Can't sleep, clown will eat me, CanOfWorms, Caniago, CanisRufus, Caponer, Carabinieri, Carodejnik, CarolGray, Catgut, Cburnett, Cera Nola, ChKa, Chainmaster, Chairman S., Champsdfw, Chanheigeorge, ChaosNil, Charlesdrakew, CharlieRCD, Chase me ladies, I'm the Cavalry, Chaucer, Chay dud, Che090572, Cherrypop09, ChildofMidnight, Chris.dillon, Christopher Parham, Chriswiki, Chuunen Baka, CieloEstrellado, Cirt, Ck lostsword, Ckatz, Classical geographer, CliffC, Cliffb, Closedmouth, Clubtransmediale, Cocytus, Codingmasters, Colipon, Collinsafag, Colors99, CommonsDelinker, Condem, Conscious, Conversion script, Corax, Corporal Tunnel, Courcelles, Coyets, Cpettauer, Crazyneeds2010, Curps, D, DAJF, DARTH SIDIOUS 2, DJ Clayworth, DLiebisch, DVD R W, DaQuirin, DabMachine, Daleydavies, Damista, DanSchultz, DanielCD, DanielRigal, Daniela Valjan, Darkildor, Darklock, Darshbegarsh, David Justin, David Kernow, David Liuzzo, Dawn Bard, Dbach, Dbenbenn, Dcandeto, DeJaVu921, Deadstar, Debresser, Deelkar, Dekisugi, Deltabeignet, Denniss, Deprifry, Der Eberswalder, Der Wolf im Wald, DerBorg, DerHexer, Dersh, Determinate, Deus Ex, Dewritech, Dhp1080, Diamondage, Diceman, Dino340, Diruhel, Discospinster, Djmutex, Dkriegls, DocWatson42, Dockingman, Doco, Docu, Dolovis, Dolphin Jedi, Domino theory, Dontworry, Dori, Doulos Christos, Download, Dr jkl, Dr.K., Dr31, DrKiernan, Drakcron, Drianhoward, Drmies, Drogo Underburrow, Dtaw2001, Duncancumming, Dunne409, EJF, ESkog, Eclecticology, Ed g2s, Edcolins, Eddideigel, Eddie6705, Edinborgarstefan, Eeekster, Eio, Ekonontheekonon, El C, El Krem, ElNuevoEinstein, Elf-friend, Eliashedberg, Eliz81, Elockid, Endurance, Engelmann15, Engunneer, Eoghan888, Epbr123, Eran of Arcadia, Erickoch, Esauvage, Esc luver, EscapingLife, Esmerálda, Esperant, Estoy Aquí, EugeneZelenko, European Girl, Evanescenceboy, Evansmith1122, Evercat, Everyking, Ewlyahoocom, Excirial, Extra999, FDT, FF2010, FJPB, Face, Feenix, Feierabend, Femto, Ferkelparade, Ferozfaza, Fertuno, FinnishDriver, Firsfron, Flammingo, Floydian, Flyguy649, Fooloomanzoo, FranceMens, Fred Bradstadt, Freddie72, Fredil Yupigo, Frenty456, FreplySpang, FritzPaulVater, Frmatt, Funandtrvl, Funnyhat, Fuse101, Future Perfect at Sunrise, Futurebird, G-41614, G. Capo, Gabbe, Gadfium, Gaius Cornelius, Galis, Galoubet, Gamer12895, Ganymead, Gbaddorf, Gensanders, Gfoley4, Ghewgill, Ghurabadeen, GiZiBoNG, Giftlite, Gilgamesh, Gilliam, Gogo Dodo, Goober-peas, Good Olfactory, Goodshope, Gplefka, GraemeL, Gralo, Grammar777g, Greatgreenwhale, Green Giant, Greenshed, Greg Godwin, Greo0, Grollp, Gronky, Ground Zero, Grover cleveland, Gryffindor, Guaka, Guanaco, Guided, Guidod, Gurch, Gustavb, Guy Peters, Gyrofrog, Gz33, Gzkn, H. Jonat, Hadal, Hadžija, Halgin, Halibutt, HamburgerRadio, Hamilton365, HangingCurve, Hanssterkendries, Happytrail, Hardouin, Harry, HarryBa, Harryboyles, Hasel, Horst, Hashman420, Hataylibo31, Hawkmc, Hayden120, Hearts of clear lake, HeikoEvermann, Heil Hellwig! Führer und Wikikanzler, Hemi1608, Henrys931, Here, Herra660, Herthaner93, Hezzard, Hhc2, Hhielscher, Hipocrite, Hmains, Homestarmy, Hornymanatee, Hthomasc, Hubschrauber729, Huckfinn1, I dream of horses, INTERRAILS, IW.HG, IZAK, Iacobus, Iamiyouareyou, Ian Pitchford, Ianb, Ide Nov, Imars, Immunize, Imperium Europeum, Improbcat, Improv, Improver 03 04, Ine12xso, Infernet, Inter16, Interlaker, Interwiki de, Inwind, IsarSteve, Ius, Ivanopolo, J.delanoy, JForget, JFreeman, JHK, JIP, JLogan, JNW, JPD, JSpung, Jackol, Jacques2, Jaellee, Jahsonic, Jake Wartenberg, Jakmcd, JamesBaine, JamieS93, Jamiegoddard4, Jammycakes, Jane McCann, Jaraalbe, Jared Preston, Jason M, Jax-wp, Jay Litman, Jaypeel, Jdavidb, Jdude2007, JeLuF, Jeff Rutsch, Jeff dean, Jeltz, JenRa, Jennica, Jeremiah, JeremyA, Jeronimo, Jgrimmer, Jiang, Jimini, JinJian, Jiy, Jketola, Jlin, Jmabel, Jmorrison230582, Jnivekk, JoanneB, Johan Magnus, John Paul Jones 34, John Paul Parks, John of Reading, JohnnyWhonny, Johnnyw, JonMoore, Jonathan.s.kt, Joowwww, Jopug, Jor, JorgeGG, Jose77, Joseph Solis in Australia, Joshlk, Jossi, Joy, Jpbowen, Jules7484, Julian Mendez, Juliancolton, Junafani, Jusdafax, Just the way, Jwnabd, Jwolf1979, Jwy, KAMiKAZOW, KJohansson, Kahn Souphanousinphone, Kanaye, Kantianer, Kappa, Karanime, Karl schroeder, KarlMathiessen, Karlos87, Karmosin, Katalaveno, Katt, Kbdank71, Kcuello, Keelm, Keith D, Keizuko, Ken Gallager, Kenpen98, Kesselring88, Ketiltrout, Killarney Daniel, Kilon22, Kim53, King of Hearts, Kingjeff, Kingpin13, Kingturtle, Knarf-bz, Koavf, Koyaanis Qatsi, KrakatoaKatie, Kresspahl, Kris12, Ksenon, Kukini, Kungfuadam, Kungming2, Kuru, Kusma, Kwamikagami, Kwertii, LAX, LUCPOL, Lahiru k, LarsHolmberg, Laurinavicius, Le Fou, Lear 21, Lectonar, Lee M, Lefterispapi, Leifern, Leuko, Levin Holtkamp, Lfh, Lhoaxt, Li-sung, Liebeaffe, Life of Riley, Liftarn, Lightdarkness, Lightmouse, Ligulem, LilHelpa, LinDrug, Ling.Nut, LitaMartin, Llywelyn, Llywrch, Loco4240, Logan, Longbow4u, Longhairandabeard, Lucki19, Luk, Luna Santin, Lupin, Luuva, Lyaios, Lycurgus, M0rph, MECU, MER-C, MGA73, MJCdetroit, MONGO, MPerel, Mac, Madchen Hoch, Madden, Madi013, Magnus Bäck, Malafaya, Malcolm, Malcolm Farmer, Malljaja, Man vyi, Manderiko, MarcelLionheart, Marcika, Marco polo, Marcsin, Marcus2, Marek69, Marekkaleja, Mariuszkalczewiak, Mark Lewis, MarkSutton, Markov Vasiliy, Markrushmore, Markussep, Marrante, Martin451, Martinwilke1980, Mathpianist93, Mattbr, Mattbuck, Matthead, Matthewedwards, Matthias Röder, Maturion, Mav, Maxboner, Maya, McDogm, Meekywiki, Meiktila, Melthamman, Menchi, Mentifisto, Merbabu, Merchbow, Mernen, Metricopolus, Meursault2004, Mevsfotw, Michael Hardy, Michael Zimmermann, Michaelm, Michaelmas1957, Michellecrisp, Midnight Green, MikeWazowski, Mild Bill Hiccup, Mileeast, Miles Pateman, Miranda, Miroslawa, Misterblues, Mjd666, Mjfv, Mkpumphrey, Mm40, Modernist, Momo san, Moncrief, Monsterxxl, Montrealais, Morn, Morwen, Mosterl23, MrBland, Ms2ger, Msikma, Mulad, MusicGeek101, Muya, Mxn, Myanw, N-true, Nachteilig, Nader, Nakon, Nathan Johnson, Natyuyyyx, NawlinWiki, Neep, NeroN BG, Nestore, Net-net, Netoholic, Neutrality, Neverquick, Newkai, Newnoise, Nick San, Nico, Nikai, Nilankapieris, Nilshennes, NisseSthlm, Niteowlneils, Njaelkies Lea, Nlu, NoSync, Nogoodfacts, Noozgroop, Norena, Nothingbutmeat, Nubzor, Nzpcmad, O Fenian, OOODDD, Ohconfucius, Okapi, Oleanna1104, Olessi, Olivier, OllieFury, Omicronpersei8, OnNYTurf, One Way Motorway, Only Red, Only green 2, OpenToppedBus, Orange Suede Sofa, OwenBlacker, OwenX, Oxymoron83, P b1999, P. S. Burton, PS2pcGAMER, PZFUN, PabDab, Pabix, Page Up, Paine Ellsworth, Pais, Paralympic, Pariiiisssian, Patrick, Patrick Sanders, Patriot 1992, Paul Benjamin Austin, Pax:Vobiscum, Pbponvho, Pcb21, Pdcook, Peaceworld111, Pedantic of Purley, Pensionero, Per Honor et Gloria, Peregrine981, Perfectmiss, Persian Poet Gal, Peter Isotalo, Peterzier, Ph89, Pharaoh of the Wizards, Philip Baird Shearer, Philip Trueman, Philippe, Phonemonkey, Picapica, Piccadilly, Piddisk, Pigsonthewing, Pilotguy, Pio Box, Piperh, Piratelooksat30some2001, Pjort, Planemo, Pokrajac, Polaron, Polli1233, Polounit, Pomte, Portia327, Postlebury, PranksterTurtle, Preaky, Princeali95, ProhibitOnions, Pschemp, Psd, Puffin, Puregenius, Putterscotch, QTCaptain, Quebec99, QueenCake, R Lowry, R sirahata, R3m0t, R9tgokunks, RA0808, RHaworth, RJN, ROGNNTUDJUU!, Rama, Randwicked, Raphie, Ravn, Raymond in der englischen Ausgabe, Rczimler, Reach Out to the Truth, Really Evil Pole, Reaper Eternal, Rebroad, Reenem, Reichsgauleiter Hansen, Reinthal, Renata, Retired username, RetiredUser2, Rettetast, Rex Germanus, RexNL, Rhcastilhos, Riana, Rich Farmbrough, Ripeugenedebs, River-phoenix-fan93, Rjh1962, Rjwilmsi, Rkt2312, RobNS, Robendo, Robert Clausen, RobertG, Robertsteadman, Rockybiggs, Roleplayer, Romuald Wróblewski, Rootlevel, Rosensteel, Rracecarr, Ruud64, Rwyeoa, Ryan, S.K., SGBailey, SH84, Sakhani, Sam Hocevar, Sam Korn, Samantha1102, Sammybhushan, Samtheboy, Samuelsen, Sander123, Sandman, Sardanaphalus, Saxifrage, Schewek, Schlammbad, Schofield797, Schuetzm, Schultheis, Sciurinæ, Scott.wheeler, Seabhcan, Sean WI, Sebastian scha., SebastianHellmann, Sebesta, Secret (renamed), Secretlondon, Seksen iki yüz kırk beş, Semlow, Serte, SeventhHell, Sevilledade, Sexyguy44, Sfdan, Shadowjams, Shayan183!, Sherool, Shibbybrohan, Shimgray, Shran, Shushruth, Siegheil, Sietse Snel, Silverhelm, SimonP, Sinep gib groot, Sir Arthur Williams, Sir James D, Skeeter08865, Skyaxx, Sligocki, Sluzzelin, Smith03, Smith2006, Smmurphy, SnapSnap, Snarius, Snoozlepet, Snowolf, Snoyes, Some jerk on the Internet, Someoneinmyheadbutit'snotme, Space Cadet, Spelled word, SpencerWilson, Spieren, Spitshine, Spongeman2007, SpookyMulder, Squarrels, Srikeit, Ssolbergj, Staeiou, Starbois, Stealman, Steez4, Stephenb, Stern, Stevegillis01, Stonemad GB, Strovolos01, Stuartrocks@gmail.com, Stuckinkiel, Styler 13, Styrofoam1994, Subtropical-man, Sundar, Sunny256, SuperHamster, Supercoop, Supersaiyanplough, Suruena, Sverdrup, Swarve, SwedBoy, Swift as an Eagle, Symane, Szylla2011, T@nn, TCTomm, TFOWR, TShilo12, TaborG, Tadas12, Takamaxa, Tapir Terrific, Tassedethe, Tatrgel, Tawker, Tbhotch, Tchaika, Teddycom, Template namespace initialisation script, Tenerer, Terfili, Texmon, Texture, The Boy With A Torn In His Side, The Thing That Should Not Be, The wub, Thedarkhawk, Theologiae, Thesunrise1983, Thewikipedian, Thialfihar, Thinem, ThompsJohn, Thucydides of Thrace, Thundernlightning, Thuresson, Tickle me, Tide rolls, TigerShark, Tim Starling, TimeClock871, Timmypees, Tke101, Tmol42, Toanke, Tobby72, Tobias Conradi, Tobibln, Toby2020, TobyJ, Tom harrison, Tommy2010, Tony Fox, Topworld International, TorLillqvist, Totti76, Tournesol, Tpbradbury, Tracyfennell, Trainspotter, TransientAircraft, TreadingWater, Treisijs, Tresiden, Tricknik, Trusilver, Tsange, Tsja, Tstitz87, Tukogbani, Tungol, Twoshirou, Tzale, Uirauna, Uk-giacomo, Ukexpat, Ulf Heinsohn, Ulric1313, Ultratomio, Unify, Urban-b, Urmas, Username4322, UtherSRG, Vadiaa, Vadmium, Van helsing, Vary, Venerable Train, Venividiwiki, Ventur, Verdadero, VeronicaPR, Verpacker Ing., Vicki Rosenzweig, Viipuri, VivaEmilyDavies, Viveiro83, Vrenator, WLU, Wadsworth57, Wasted Time R, Waterguy, Wavelength, Wayward, Weasel, Weissmann, Wes!, Westee, Weygre, Wfgiuliano, Wheatley000, Whiskasisakitten, WhisperToMe, Whispering, Wik, WikHead, WikiLaurent, Wikid77, Wikilikedl, Wikinaut, Wikiniki, WildlifeAnalysis, Will-i-am558, William Avery, Wilt, Winchelsea, Wireless Keyboard, Wisden17, Wjw0111, WojPob, Wolfdog, WolframSiever, Woohookitty, Worldglobal, Wow Scotland !, Wtmitchell, Wzwz, X201, Xakaxunknownx, Xakon, Xndr, Yaan, YacozA, YellowMonkey, Yintan, Yitzhak, Yngvadottir, Yoninah, YourEyesOnly, Yurigerhard, ZS, ZapThunderstrike, Zbcc, Zerogeewhiz, Ziko, Zocky, Zoe, Zonk43, Zundark, Zzuuzz, Ã◆£, Директор, קוראפ, 3110 anonymous edits

Berlin S-Bahn *Source*: http://en.wikipedia.org/w/index.php?title=Berlin_S-Bahn *Contributors*: AalanTG, Acvb, Ajcoxuk, Alai, Ali@gwc.org.uk, Angr, AnonMoos, Anorak2, ArtVandelay13, Artpol, BajanZindy, Barliner, Bermicourt, BetaCentauri, Bhludzin, Bjarki S, Bobrayner, Bryan Derksen, Captndelta, Chris j wood, Chumwa, Cmdrjameson, Cod, CommonsDelinker, Cosmobird, Cs-wolves, Cvieg, Czyrko, Daduzi, DerBorg, DisillusionedBitterAndKnackered, Doco, Emn884, Eraserhead1, Gaius Cornelius, Gamgee, Guirro, Gwernol, Hhielscher, Huaiwei, Inwind, IsarSteve, JGHowes, JNZ, Jacoplane, Jfruh, Joshua, Jpbowen, KTo288, Keatinga, Kelisi, Kildor, Kingjeff, Kv75, Larry V, Lear 21, Lightmouse, MFantti, Mackensen, Marianian, Marj Tiefert, Mattbr, Max rspct, Metropolitan, Morwen, Mr WR, Not-just-yeti, Ozdaren, Pedantic of Purley, PeterR, Pokrajac, RedWordSmith, Rheintalfan, Rich Farmbrough, Rjwilmsi, Rlquall, Secretlondon, Slambo, Solarapex, Squibman, Starbois, Stereorock, Synchronism, Tim!, Timothyhouse1, Tkynerd, Tobibln, Typochimp, Uaxuctum, Uzdzislaw, Valenciano, We hope, Wik, Yngvadottir, 76 anonymous edits

Commuter rail *Source*: http://en.wikipedia.org/w/index.php?title=Commuter_rail *Contributors*: A.arvind.arasu, Adam78, AdamW, Admiral Norton, Advanced, Ainlina, AlMac, Alan Liefting, Alan.ca, Alanmak, Albertocchaves, Amire80, AndrewHowse, Andrewpmk, Andycjp, Angusmclellan, Armigo, Arsenikk, ArtVandelay13, AvicAWB, BD2412, Badger151, Bean49,

Benandorsqueaks, Berendale2, Bermicourt, Bg007, Bigdumbdinosaur, Caltas, Camerong, Cassowary, Central Data Bank, Cfaerber, CommonsDelinker, Conversion script, Covalent, Crd721, Crissov, DAJF, Darkcore, David.Monniaux, Dbehr, Dellarb, Dillan.Murray, Dolovis, Drhlajos, Dual Freq, Edward, Eelamstylez77, Eisfbnore, EoGuy, Eraserhead1, Erianna, Escape Orbit, Ewlyahoocom, Fairlyoddparents1234, Filip95, Frietjes, Gaius Cornelius, Gidonb, GreekRailFan, Gumbogumbo, Harryboyles, Hazmat2, Henning Makholm, Hokachung, Htonl, Inwind, J Hazard, J.delanoy, Jabez Stone, Jim.henderson, Jonahgreenthal, Joseph Solis in Australia, Joshua, Joy, Jsfouche, Kenrick95, Km9, KuwarOnline, Lacrimosus, Larry V, Leinsd, LilHelpa, Loco830, Loodog, Loqu, Lucaf1, Mahjongg, MattEHurst, MickMacNee, Mikiqex, MissionInn.Jim, Moebiusuibeom-en, Mog, Montemonte, Montrealais, Mouchoir le Souris, Moyabrit, Mpotse, MrLincoln, Mrezee, Neckintosh, Neutrality, Nixer, NoAccount, Nopira, Oknazevad, Olivier, Optimist on the run, Orbitalwow, Osanjose, Patrick, Peter Horn, Picapica, Pietrow, Plastikspork, Playclever, Plyhmrp, Radagast, Randwicked, Raphaelmak, Raviytedz, Rjwilmsi, Rovibroni, Rsrikanth05, S.S.Exp.Hashimoto, SPUI, Saruwine, Sebastian scha., Sebwite, Shy Guy Gunzel, Sirap bandung, Skgxt2, Slambo, Smack, Speciate, Steve chiu, Sushiya, Tawker, Themanilaxperience, Threecharlie, TiffaF, Tompw, Toytoy, Vaoverland, Vmenkov, We hope, Welshleprechaun, WinC78, Wongm, Woohookitty, Wuhwuzdat, Zoney, 160 anonymous edits

East Berlin *Source*: http://en.wikipedia.org/w/index.php?title=East_Berlin *Contributors*: 52 Pickup, Adam Carr, Agathoclea, AlbertR, Amandajm, Analgape, Andre Engels, AndreasJS, Anorak2, Asgardur, Averell23, Barliner, Beland, Binabik80, Brianhenke, Cheong Kok Chun, Clerkwh, Cobaltcigs, CommonsDelinker, Conversion script, Cvieg, Den fjättrade ankan, DerBorg, Deror avi, Doco, Domino theory, EGroup, Enchanter, Fkehar, ForUsByUs, Francs2000, Geneb1955, Glen, Golbez, Good Olfactory, Gr8opinionater, Guaka, Happyme22, Hhielscher, Ians, Japanese Searobin, Jatder, Jfruh, Jolove55, Kabri, Kay Körner, Ketiltrout, Kjet, Kkin, Komischn, Kusma, LHOON, Laurinavicius, Lectonar, Lightmouse, Madden, Maelnuneb, Magioladitis, Magnus.de, Mic, Michael Hardy, Milantoin, Morwen, Nunquam Dormio, OOODDD, Olivier, Orioane, OwenBlacker, Paul Benjamin Austin, Poppy, ProhibitOnions, RedHillian, S-J-S-F-M-W, SH84, Salgueiro, Sandman, SeNeKa, Sfan00 IMG, Sir James D, Slawojarek, Slleong, Stemonitis, Superzohar, Susvolans, Tdowling, The Beagle, The very model of a minor general, Twthmoses, Ukexpat, Ulf Heinsohn, Unyoyega, Uvaduck, Valip, Voyageur, Wiggy!, WojPob, Ype, Zakuragi, Δ, 72 anonymous edits

Friedrichshain *Source*: http://en.wikipedia.org/w/index.php?title=Friedrichshain *Contributors*: Albrecht Conz, AndreHuppertz, Andreas Philopater, Angr, Arsene, Averell23, Berlininlondon, BillFlis, CalJW, Chris the speller, DerBorg, Dgwohu, Dr. Blofeld, Endurance, Gaius Cornelius, Gilliam, Guidod, Harmat, Hfaustus, Hmains, IsarSteve, Kwiki, Little Savage, Manix2000, Michael Hardy, Mikyle, Ohconfucius, Phonemonkey, ProhibitOnions, Rabenkind, Redrose64, Starbois, Superzohar, TIY, Valenciano, 39 anonymous edits

Friedrichshain-Kreuzberg *Source*: http://en.wikipedia.org/w/index.php?title=Friedrichshain-Kreuzberg *Contributors*: AndreHuppertz, Berlin 11011, Bluap, CalJW, DK4, Der Ausländer, DerBorg, Dewritech, Dravecky, Emijrp, INkubusse, Icestorm815, Lear 21, Puregenius, Sam916, Sandman, Scott.wheeler, TIY, Valenciano, Whatcanbrowndo, 33 anonymous edits

Rummelsburg *Source*: http://en.wikipedia.org/w/index.php?title=Rummelsburg *Contributors*: Angr, Bamyers99, Chris the speller, DerBorg, Ohconfucius, Olessi, Wikipelli, 1 anonymous edits

Berlin Westkreuz station *Source*: http://en.wikipedia.org/w/index.php?title=Berlin_Westkreuz_station *Contributors*: AalanTG, Bermicourt, DerBorg, Grahamec, Jef Kazak, Jumbo1435, Slambo, 2 anonymous edits

Berlin Stadtbahn *Source*: http://en.wikipedia.org/w/index.php?title=Berlin_Stadtbahn *Contributors*: AalanTG, Ajcoxuk, ArtVandelay13, Avicennasis, Axpde, Bermicourt, CommonsDelinker, Darklilac, Delirium, DerBorg, Doco, Dseifert, Eisfbnore, Erik9, Grahamec, Guidod, InspectorTiger, Lightmouse, Ling.Nut, Ms2ger, Neilbeach, Ospalh, Peter Horn, Philip Baird Shearer, Pyrotec, RHB, Rjwilmsi, Starbois, Thiseye, Tony1, Wlindley, Yngvadottir, 9 anonymous edits

Berlin Ringbahn *Source*: http://en.wikipedia.org/w/index.php?title=Berlin_Ringbahn *Contributors*: AalanTG, Axpde, Bermicourt, Captndelta, DerBorg, Doco, Erik9, Grahamec, Hinto, Miguel Andrade, Plasticup, Sladen, Starbois, Welsh, Yngvadottir, 8 anonymous edits

Berlin Wall *Source*: http://en.wikipedia.org/w/index.php?title=Berlin_Wall *Contributors*: -Majestic-, -jkb-, .:Ajvol:., 041744, 1.21 jigwatts, 159753, 172, 1j1z2, 2012WorldPeace, 23skidoo, 28421u2232nfenfcenc, 4twenty42o, A gx7, ABF, AP1787, Abaris, Ackatsis, Acntx, Acroterion, Adam78, Adam850, Addshore, AdjustShift, Ae7flux, Afc, Agathman, Agathoclea, Ahoerstemeier, Aicchalmers, Aigues-Mortes, Aillema, Airplaneman, Aivazovsky, Akamad, Akhristov, Akilduff, Alan Liefting, Alandeus, Alansohn, Aldis90, Ale jrb, Aleenf1, AlexBRA, AlexRat, AlexanderKaras, Alexandra lb, Alexius08, AlexiusHoratius, Alexkiller17, Ali@gwc.org.uk, AlienZen, Allstarecho, Altenmann, Altes, Amakuru, Amalthea, AmericanZombie, Ames5929, AmiDaniel, Analcheese, Ancheta Wis, Andareed, Andonic, Andrea105, Andrew Levine, AndrewN, Andrewpmk, Andrewsteven, Andy Marchbanks, AngelOfSadness, Angela, Anger22, Angr, Anna manana, AnnaFrance, AnonMoos, Anorak2, Antandrus, Anthony Appleyard, Anti-Nationalist, Antonio Lopez, Arcadie, Archanamiya, Artaxiad, Arthena, Artiste-extraordinaire, Arwel Parry, Asdfasdfasdfasdf1, Atavi, Atfcbrad, Athaenara, AtheWeatherman, Atif.t2, AtomBoy, AuburnPilot, Avicennasis, Avoided, AxelBoldt, Axeman89, Ayla, AzaToth, B. Fairbairn, B.d.mills, BBrad31, BLan, BLueFiSH.as, BSVulturis, Bact, Badger 990, Badger 991, Badger 992, BadgerMoose, Badgernet, Balcer, Bansal123, Barjag, BaronLarf, Bastin, Bayerischermann, Beefalo, Before My Ken, Belgiumboy108, Belovedfreak, Belugaperson, Ben Jos, Bencourtney, Bender235, Benhaldenby, Berlina-baer-en, Betacommand, Bettia, Bibi999, Big red circle, BigHairRef, Bigfatty1, Bkwillwm, Blair Bonnett, Blanchardb, Bleh999, Blehfu, Blisco, Blowski, Blueshirts, Bobak, Bobblewik, Bobby1011, Bobe13, Bobo192, Boccobrock, Bogey97, Bongwarrior, Boothinator, Bouktin, Brandmeister, Brandmeister (old), Briaboru, Brian the Editor, Britmax, Bruce1ee, Brushwell, Brutaldeluxe, Brycen, Bsimmons666, Buckshot06, Bukubku, Burntsauce, Bushido93, C.Fred, C.J. Griffin, CIreland, CMacMillan, Calbaer, Calmer Waters, Calor, Caltas, CalumH93, CambridgeBayWeather, Can't sleep, clown will eat me, CanadianLinuxUser, Candicechristiane, Candidhq, Captain ranga101, Captain-n00dle, Carafe, Carajou, CardinalDan, Carmelbuck, Catsrock1515, Cavan, Cfaerber, Cgingold, Chameleon, Charles Edwin Shipp, Charlesdrakew, Chcknwnm, Cheungpat, Chim43ra, Chochopk, Chriscoolc, Chrishmt0423, Christian75, Chrysalis, Chuckie Norris, Chunky Rice, Civil Engineer III, Ckatz, Clasqm, Classical geographer, Cliffd62, Clindberg, Clint0601, Closedmouth, CloudNine, CoderGnome, Codyman, Colipon, Colonies Chris, Commander Keane, Commonbrick, CommonsDelinker, Computerdan000, ComradeAxe, ConnorHemme, Conscious, Constanz, Conversion script, Cooluncle55, Corker1, Cosmic Latte, Courcelles, Coyets, Crazycomputers, Crevox, Cripipper, Crohnie, Cubfanpgh, Curps, Cutcher, Cvieg, Cybjorg, CzarB, DARTH SIDIOUS 2, DD2K, DMacks, DO'Neil, DVD R W, Da chucky, Dac04, DagosNavy, Dakusan, DamienLaughton, Daniel, Daniel Case, Daniel J. Leivick, Daniel.Cardenas, Danielsh, Danlaycock, Danleo, Danno uk, Dark Mage, Datrain329, Davephil6969, Davewho2, David Ruffle, DavidLevinson, Davitasaun, Dazzla, Dc student2011, Dchall1, Dcosman, DeLarge, DeadEyeArrow, Deagle AP, Deathbythebill, Debeo Morium, Debresser, December21st2012Freak, Deckiller, Demiccoli, Demiurge, Den fjättrade ankan, Denki, DennyColt, Deor, DerBorg, DerHexer, Derekbd, DevastatorIIC, Dewritech, Dhlstrm, Dhyeatts, Diceman, Dickgomer, Digwuren, Dina, DirkvdM, Discospinster, Djmutex, Dmkrantz, Dmytro, Doc glasgow, DogcatcherDrew, Dolovis, Dominus, DonaldDuck, Donarreiskoffer, Dondelinger, Donfbreed, Donkeyman00001111, Doopdoop, Dothefandango, Dozer5274, Dr Santa, Dr. Blofeld, Dr. Dan, Dragon325455, Drbreznjev, Drewbaseball93, Dudeman5685, Dunbar14, Dureo, Durova, Dv82matt, Dylanator190, Dynamax, Dysepsion, Dysprosia, DéRahier, ESkog, EXODUS7, Eaefremov, EagleOne, Eaglestorm, Ec8or, Eclecticology, Ecoleetage, Ed Poor, Edcolins, Edit krazy, Editor510, Edivorce, Edmundbaird, Edward, Edwy, Egkauston, Ehheh, Ei2g, Ekanning, Elephant53, Eli Wim, Elipongo, Elitehax, Eliz81, Elizard16, Elkman, Elockid, Emc2, Emhoo, Enchanter, Entropy, Enviroboy, Epbr123, ErdemTuzun, Erebus Morgaine, Erekrose, Ericdn, ErikNY, ErikRomijn, ErrantX, Erujiu12, EscapingLife, Esmerálda, Evangelion9490, Everyking, Evil Monkey, Evil saltine, Evlekis, EvocativeIntrigue, Excirial, F, F-402, F141998, FDT, FF2010, Falcon8765, Fan-1967, Fastily, Fatalerrorrage, Fatgirl99, Favonian, Feitclub, Ferzapata, Fethroesforia, Fieldday-sunday, Fighting for Justice, FisherQueen, FitzColinGerald, Flatterworld, Fleminra, Floydian, Flyingember, Footix, Formeruser-81, Franamax, FrancoGG, FranksValli, Freakadorkoid, Fredrik, Freerick, Frietjes, Friginator, Frogman123, Fromgermany, Fubu01, FullMetal Falcon, Funnyhat, Fursday, Future Perfect at Sunrise, Fuzheado, Fvasconcellos, Fys, G-Man, GGreeneVa, GSPT1, Gabbe, Gabe7285, Gabrielepiccinnu, Gail, Gaius Cornelius, Gangsta1121, Garethhamilton, Gary King, GateKeeper, Gatfish, Gbrandt, Gedca, GenericBob, Geordie2.0, GeorgeOrr, Gerda Arendt, GermanyInfo, Ghbvgy, Gilliam, Gillyweed, Gimboid13, Ginsengbomb, Gladmax, Gmaxwell, GnarlyLikeWhoa, Goatasaur, Gogo Dodo, GoldCorrie, Good Olfactory, GraemeL, Grafen, Graham87, Gralo, GreenReaper, Grifter72, Grinder0-0, GrooveDog, Grubel, Gruntled Student, Gtg204y, Gtstricky, Guayl, Gugganij, Gumpus, Gurch, Gurubrahma, Guusbosman, Gwb16, Gwernol, Gyurika, H1e2n3r4y5, H2g2bob, Hadal, Haemo, Halda, HalfShadow, Halgin, HamburgerRadio, HangingCurve, HappyJake, Happyme22, Happytrail, Hardmoose1, HarlandQPitt, Havantshire, Hayvac, Hbent, Hdt83, Heezy, HeinzzzderMannn, Hemmingsen, Heracles31, HerkusMonte, Himarty, Hjholland182, Hmains, Hollister1234, Horsebrutality, Howcheng, Huggle, Hullie, I am neuron, IIVQ, ILike2BeAnonymous, IRP, Ianb, Ians, Icancounttoten12345678910, Icestorm815, Icurok, Ida Shaw, Ignoramibus, Igoldste, Iluvcapra, Immunize, ImperatorExercitus, Ingolfson, Insanity Incarnate, Intelligentsium, Into The Fray, Iridescent, Isangaft220, IsarSteve, Isecore, Islander, IstvanWolf, Itsabouttime, Itsfullofstars, Itsmine, Ivanko15, Iwalters, Iwearshoess, Ixfd64, J 1982, J.delanoy, JAJ, JALockhart, JForget, JFreeman, JSpung, Jackol, Jake06s, Jakemurray23, JamesBWatson, Jamesooders, Janko, Jao, Jaraalbe, Jared Preston, Jarsyl, Jasongill, Jauerback, Jaxl, Jayjg, Jaypee1, Jayron32, Jbalint, Jbernardo92, Jcobrien, Jdlh, Jebtinker, Jeff G., Jelligraze, Jengod, Jeronimo, Jfruh, Jfwilsonmd, Jiddisch, Jimbowimbo2, Jivecat, Jj137, Jlascar, Jmabel, Jmlk17, Jnorton7558, JoanneB, John, John Paul Parks, John254, JohnWittle, Jojhutton, Jonadab, Jonathunder, Joseph Solis in Australia, Josh manton 16, Jossi, Joy, Joyous!, Jpatokal, Jpbowen, Jpgordon, Jsc83, Jtanadi, JulianKurz, Jumbuck, Jusdafax, Jusjih, JustPhil, Justins123, KGasso, Kai81, Kaisershatner, Kamditch, Kanags, Kansan, Kar98, Katharineamy, Katydidit, Kbdank71, Kcordina, Ke4roh, Keenan Pepper, Kehrbykid, Keilana, KeithTyler, Kelly Martin, Kevin B12, Kevin j, Kewkid, Khazar, Killgor66, King of Hearts, Kingal86, Kingjeff, Kingpin13, Kingturtle, Kintetsubuffalo, Kirill Lokshin, Kismetmagic, Kitch, Kkcrackhead12, Knowledge33, Knverma, Koavf, Kobalt08, Konstantin, Koobaxion, Korean77777, Kornfan71, Kosebamse, KoshVorlon, Kukorelli, Kumar Appaiah, Kumioko, Kungfuadam, Kuraikatara, Kuru, Kurykh, Kusma, Kvasir, Kwamikagami, Kwekwc, Kwertii, KyraVixen, Kzero22, L Kensington, LARS, LFaraone, LOL, LaMenta3, Lacrimosus, Ladyasays, Laenir, Lahiru k, Landon1980, Lankiveil, Larrymcp, Laudaka, Lazar Taxon, Ld, Le sacre, Lectonar, Leithp, LeonardoRob0t, Leszek Jańczuk, Leuko, Leutha, Levineps, Lfh, Lightdarkness, Lightmouse, Ligulem, Lilatl217, Lily15, Limabeans, Linguoboy, Liquidjp, Lisaviktorya, Llort, Llywelyn, Longhairandabeard, Looxix, LordAmeth, Loren36, Lothar von Richthofen, Lradrama, LugxXx, Luke Bales, Luna Santin, Lupo, Lurlock, Lyricmac, M.thoriyan, M1ss1ontomars2k4, M2MM4M, MADaboutforests, MBK004, MDM, MER-C, MFWS, MONGO, Maccers, Macintosh User, Madmaddy27, Madoreck, Mailer diablo, MajorStovall, Majorclanger, Malcolm Farmer, Malinaccier, Malkinann, Mamalala, Manormadman, Manvegas, Marcinx, Marek69, Martarius, MasonBanks, Master200021, Materialscientist, Mathfan, Matthead, Matthew Yeager, Mattymatts123, Mauls, Mav, Max rspct, Maxamegalon2000, Maxell222, Maxim, Maximus Rex, Maxl, Maxonfire, Mbc362, Mboverload, Mbz1, Mc6809e, Mcburger77, Mccleskeygenius21, Mdrejhon, Meaghan, Meekywiki, Melancholia i, Membrstff455677, Mentifisto, Mephistophelian, Merlion444, Mernen, Metalicballs123, MetsFan76, Meursault2004, Mfield, Mgiganteus1, Michael C Price, Michael Hardy, Michael Zimmermann, Microcuts92, Mike Rosoft, Mikeo, Milkyface, Milo99, Mimzy1990, Mintguy, Mitchman93, Mjpieters, Mlm42, Mmortz, Mnh123, Moasir, Modest Genius, Moogyboy, Mormegil, Morwen, Mosedschurte, Mr. Lefty, Mr. Yooper, Mrannen1958, Mranostay, Mschamis, Mug1234, Muspud2, My76Strat, Mygerardromance, Mygocarp, Myphoneisretro, Mysidia, N2e, NHRHS2010, NOyieldingCODE, Naddy, Nagging, Nakon, Nakosomo, Name? I have no name., Narcissus14, Naruttebayo, Natahn, NativeForeigner, Natrajdr, NawlinWiki, NellieBly, Netwrox, Neuro, Nev1, Neverquick, Nevfennas, NewEnglandYankee, Newone, Newtaste, Nichiyoubi, Nicholas.Tan, Nick, NickW557, Nickel Chromo, Nickname3000, Nickptar, Nicolou101, NigelR, Nightscream, NinjaCharlie, Nitrowolf, Nivix, Nlu, Nneonneo, Noctibus, NodnarbLlad, Normaldesign, NorthernThunder, NorwegianBlue, Noseybosey, Not telling, Nottheking, Nsaa, Nturner14, Nubiatech, Nuggetboy,

Nugneant, Nunocordeiro, Nunquam Dormio, ObfuscatePenguin, Oblivious, Odie5533, Ohconfucius, Ohya23, OlEnglish, Olivier, Omgroflol, Omphaloscope, One, Onebravemonkey, Orbmc101, OregonD00d, Orpheus, OwenX, Oxymoron83, POTA, PRRfan, Pagrashtak, Pako, Paranomia, Parkwells, Pat Payne, Patrick, Paul Benjamin Austin, Paul-L, Pauldipa, PaulinSaudi, Pavel Vozenilek, Paxsimius, Pdwerryh, PedanticallySpeaking, PenguiN42, Penhollow, Per1892, Persian Poet Gal, Peruvianllama, Peter Deer, Peter Isotalo, Petersent, Peterzier, Petri Krohn, Pevernagie, Pfainuk, Pgk, Pharaoh of the Wizards, Phil Boswell, Phil Sandifer, Philannetta, Philip Trueman, Phlegat, PhnomPencil, Phrenq, Piedaemon, Pigsonthewing, Pinethicket, Piotrus, Pippin25, Plumpy, Pmanderson, Pol430, Populus, Porky149, Postdlf, Ppk01, Printer258965321450, Private Pilot, ProEditor911, Prolog, Pschemp, Psternenberg, Psy guy, Publicus, Puppets2007, Pvnugteren, Pyrospirit, QTCaptain, Quae legit, Qui1che, Quintote, Qxz, R, R Lowry, R3ap3R, RFBailey, RJaguar3, RM21, Radagast, RanEagle, Ranaenc, RandomAct, Rangermike, Ranmaliebe, Raven4x4x, Raymond, Razorflame, Rbakker99, Rbanzai, Rcktmanil, Rd232, Reach Out to the Truth, Readro, Rebecca, Recognizance, Red3334, RedHillian, Redrey, Redthoreau, Reinoutr, Rettetast, RexNL, Reywas92, Rezashah4, Riana, Rich Farmbrough, Richard Arthur Norton (1958-), Richard Weil, Rick Block, Rickyjames, Rikanderson, Rjwilmsi, Rlove, Rmhermen, Rnt20, Robbyjob, RobertG, Robertpdot, Robertvan1, Robin klein, Rockero, Rockpocket, Rodhullandemu, Roleplayer, Romanm, Ronhjones, Ropers, Routlee, RoyBoy, Rrburke, Rrmsjp, Rsrikanth05, Russavia, Ruy Lopez, Rw rynerson, Ryan Postlethwaite, RyanCross, Ryulong, SGGH, Saint-Paddy, Saintswithin, SallyBoseman, Salmanazar, Saltdeanmark, Samtheboy, Samugariya, SanFranEditor, Sandmountainslim, Sango123, Sarbear05, Savidan, Scapler, SchfiftyThree, Schoelle, Schumaier, SchuminWeb, Schzmo, Scientizzle, Sciurinæ, Scocar27, Scog, Scoops, Scott14, Seaphoto, Sebastian scha., Sebleblanc, Secret (renamed), Seek, Semperf, Seoulmate1, Sequitur, Seraphim, Severa, Shadowjams, Shalom Yechiel, Shanes, Sheephead, Shen, Shenme, Sherurcij, Shimgray, Shirulashem, Shoeofdeath, Shreyasjoshis, Siggydude, Sigma 7, Signalhead, Silversmith, Sir Stanley, Sir chris 2, Siroxo, Sjakkalle, Skäpperöd, Sladen, SlimVirgin, Slleong, Sluzzelin, Sm8900, Smalljim, Smerus, Smilesfozwood, Smith2006, Smjg, Smsarmad, Smurfmeister, SnareDrumKneeCaps, Snowolf, Soap, Soccerdude265, Softlavender, Solipsist, Soman, Some jerk on the Internet, SoxSweepAgain, Spanish lullaby, Special-T, Speedboy Salesman, Spellcast, Spencer, Spitfire, Splash, Spongefrog, St07, Staka, Stamatis, StanleyIrwell, Starbois, Stargoat, StaticGull, Steel, Stefanomione, Steinsky, Stephen Burnett, Stephenb, Steve2011, SteveSims, Steven Zhang, Stevenj, Stompin' Tom, Stor stark7, StradivariusTV, Stumps, Stupefaction, Suffusion of Yellow, Sully, SunCreator, Sunderland06, Sunnyvale guy, Supaguyfly37, Superchocholatebear, Superworms, Superzohar, Susfele, Svenman, Sverdrup, Svick, Swarve, SwatSource, Swiftblade21, Sylent, Syrthiss, System123456789, TBadger, TDC, TPK, TTE, TYRXrus, Tabletop, Talonxpool, Tamas Szabo, Tamfang, Tanvir Ahmmed, Tariqabjotu, Tarret, Taw, Tawker, Tbhotch, Tea with toast, TeaDrinker, Techman224, Teles, Tellyaddict, Tempodivalse, Terence22205, TharangaJ, The Dark, The Egyptian Liberal, The Giant Puffin, The High Fin Sperm Whale, The Hoffmeister, The Ink Daddy!, The Thing That Should Not Be, The ed17, The undertow, TheBY, TheLamprey, Thebiguy, Theda, Themightyquill, Thermaland, Thewikipedian, Thingg, Thingnus, Think777, Thomas Blomberg, Thomas81, Thryduulf, Thumperward, Tiddly Tom, Tide rolls, TigerShark, Tiggerhop, TimElessness, Timmeh, Tins128, Tintero, Tiptoety, Tkotsch, Todowd, Tokus, Tom servo, TomGaines, Tombomp, Tommy2010, Tony1, TonyTheTiger, Tpbradbury, Tr33zon, Trainspotter, Trakesht, Trappist the monk, Tremblay, Trevdna, Trisar, Triwbe, Trust Is All You Need, Tuck1234, Tulandro, Turian, Tutmosis, Tvtrojan11, Twang, Tyrenius, Tysto, Tytrain, UkPaolo, Ulf Heinsohn, Ulric1313, Uncle Dick, Unify, Unschool, Urban XII, Ustye, Utcursch, Valip, Van helsing, Vancouverguy, Vanished user 03, Vanished user 39948282, Varlaam, Versus22, Verygood4u, Vess345, Vicarious, Vicki Rosenzweig, Victorgrigas, Vidor, Vignaux, Vints, Violetriga, Vladmirfish, Voyagerfan5761, Vrubel's Demons, Vulture, W. B. Wilson, WPArgentarius, Wa2ise, Wafulz, Warofdreams, Wasell, Watch37264, Watcharakorn, Wavelength, Werdan7, Wertyasman, West Brom 4ever, Westee, WhisperToMe, Wiggster, Wiki alf, WikiBrown, Wikibofh, Wikieditoroftoday, WikipedianMarlith, Wikipelli, William Avery, Willit63, Wimt, WishingForever, Witan, Wknight94, Wolskinj, Woohookitty, Wtdoor, Wtfunkymonkey, Wuhwuzdat, Wulf Isebrand, X3210, XXChaPiiXx, XerKibard, XwrappedXinXplasticXgirlX, Xymmax, Yaan, Yahel Guhan, Yahibemu, Yamamoto Ichiro, Yankeedoodle, Yann, YellowFives, Yogi de, Yonatan, Yuckfoo, Z.E.R.O., Zach mp, Zamkudi, ZapThunderstrike, Zaphraud, Zapvet, Zazou, Zepher25, ZimZalaBim, Zimmi5959, Zoonerd, Zvn, ²¹², , 3279 anonymous edits

West Berlin *Source*: http://en.wikipedia.org/w/index.php?title=West_Berlin *Contributors*: 213.84.75.xxx, 52 Pickup, 9014user, AlbertR, Alexwcovington, Anorak2, Antodav2007, BillFlis, Boffin, Boothy443, Bowlhover, Buzz, Cholling, Chris Roy, Chronodm, Ckatz, Clngre, Complex01, Conversion script, Courcelles, Cybercobra, David Gerard, Deb, DerBorg, Deror avi, Dewritech, Doco, Domino theory, DrDaveHPP, Eisenringtheo, Ekem, Epbr123, Flandes, Fuzzy Logic, Gaius Cornelius, Gamgee, Gauss, Gcapp1959, Good Olfactory, Goustien, Gr8opinionater, Graham87, Hamleteer, HangingCurve, Inter, Irakli D, J.J. Popplewick, JNZ, Jao, Jfruh, Jmorrison230582, Joey80, John, Joseph Solis in Australia, Jung dalglish, Kingturtle, Kross, Känsterle, Lacrimosus, Lightmouse, Lockesdonkey, Lothar von Richthofen, Lowellian, MER-C, MJCdetroit, MaGioZal, Malljaja, MartinZwirlein, Mauls, Maximaximax, Melchoir, Mic, Michael Hardy, Mnmazur, Morwen, MrDolomite, MusicGeek101, Mustapilvi, N Yo FACE, N328KF, NSK Nikolaos S. Karastathis, Narcosis17, Ndteegarden, Northumbrian, Ochkarik, Olegwiki, Olliemoo, Open2universe, OwenBlacker, Pail, Palica, Parsleyij, Paul Benjamin Austin, Paul-L, Peregrine981, Phase1, Philip Baird Shearer, Pichpich, Poppy, ProhibitOnions, Prumpf, Quiensabe, R'n'B, RFBailey, RMG, Radagast83, Rebroad, Rich Farmbrough, Rspeer, Rw rynerson, SGBailey, Sandman, Sannse, SeNeKa, Shadowjams, Slawojarek, Slleong, Snake bgd, SnappingTurtle, Snoyes, Susvolans, Sverdrup, Teotocopulos, Terfili, Thomprod, Tpbradbury, Tsja, Twthmoses, Ulf Heinsohn, Uvaduck, Valip, Westee, Wheatley000, Wl219, Wlamb76, WojPob, Wongm, Woohookitty, Yngvadottir, Zmx15, 193 anonymous edits

German reunification *Source*: http://en.wikipedia.org/w/index.php?title=German_reunification *Contributors*: 159753, Ahoerstemeier, AjaxSmack, Akamad, Alai, Algont, Amikeco, Andreas D.C., Andrelvis, Andres, Andy Marchbanks, Angel ivanov angelov, Angr, Antonio Basto, Arakunem, Arbor, Aris Katsaris, ArvedJoegr, Ashley Y, Austrian, Axeman89, Axpde, Axt, Azate, B.d.mills, BLueFiSH.as, Baldhur, Bender235, Benson85, Bernhardsulzer, Bevo74, Bezapt, Big Adamsky, BigFatBuddha, BilCat, Bletch, Blinder Seher, Bobblewik, Bobo192, Boson, Brandon5485, Brazzy, Britannicus, Brw12, Bry9000, Bukubku, Clreland, Caleiva, Caliscope, Chochopk, Chris Roy, Chris the speller, Clamum, Clarkefreak, Colchicum, CommonsDelinker, Conversion script, Cpl Syx, Cush, Cvieg, DARTH SIDIOUS 2, Dale Arnett, Damczyk, David Gerard, Davideskimo, Deissi, Demodave, Denniss, Devatipan, Diberri, Doco, Donarreiskoffer, DorisAntony, Dr.K., Dreadstar, DropDeadGorgias, Dub8lad1, Dynam1te3, Dynzmoar, Dúnadan, El C, Elbruz Bey, Eliz81, Elizard16, Estoy Aquí, Everyking, Explicit, Fidian, Finduilas 09, Freerick, Fryede, Funnyhat, GCarty, Gabbe, Gaius Cornelius, Gemeinsam, Geo-Loge, GeorgeLouis, Get-back-world-respect, Gkrupa30, Good Olfactory, Graham87, GregAsche, Grm wnr, Grondemar, Hall Monitor, HangingCurve, Heidimekawi, HeikoEvermann, Heinzmeier, Hoziron, Hylaride, INXS-Girl, Icseaturtles, Imladros, Immunize, Indon, Ingolfson, Ingsoc, InverseHypercube, Ithanyx, ItsZippy, Ixfd64, Jak123, Jannizz, Jao, Jaraalbe, Jason M, Jeffwang, Jfruh, Jllatimer, Jmorrison230582, Jnestorius, John Holmes II, JohnnyKenCam, Johnxxx9, Joopercoopers, Joseph Solis in Australia, Jossi, Jovmer45, Jvhertum, Kaare, Keith H., Kelisi, Kinst, Kintetsubuffalo, Kitch, Knarf-bz, Knot8head, Kojozone, Kpalion, KristoferM, Kummi, Kupredu, LOL, Laca, Lacrimosus, Lbunker, Lear 21, LeedsKing, Leushenko, Lexicon, Lightmouse, Lowellian, MaartenVidal, Malfidus, Manuel Anastácio, Marcoabcde, Materialscientist, Mathew5000, Matthead, Mav, Mbinebri, Mentifisto, Merbabu, Meursault2004, Michael Hardy, Michael Zimmermann, Miranche, Mistakefinder, Mjb, Mmunger, Molobo, Mstuomel, N328KF, Nakon, Narcosis17, Neil916, Nero the second, Nick Number, Nickshanks, Nicolasdz, Nightstallion, Nmiller22222222, Nsaa, Nunocordeiro, Obamafan70, Obersachse, Offkilter, Ohconfucius, Okj579, Oliphaunt, Olivier, PFHLai, Patrick, Paul Benjamin Austin, Pavel Vozenilek, Pearle, Pepper, Peter Grey, Petri Krohn, Phaedriel, Philip Baird Shearer, Pippu d'Angelo, PittBill, Pkmink, Pointer1, Poli, Prumpf, PuzzletChung, Qb229, R Lowry, Ran, Random user 8384993, Randwicked, Ratiocinate, Rd232, Reedy, Reinoutr, Repkow, Rich Farmbrough, Rjensen, Rjwilmsi, Rmhermen, Roadrunner, Robin klein, Routlee, RoyBoy, Russavia, Rw rynerson, SDY, Sarahstern, Schnauzendorf, SciCorrector, Scip, Scipius, Secret (renamed), Sevvvy, Skeezix1000, Skäpperöd, Slgrandson, Slowking Man, Sm8900, Snowdog, Split, Spondoolicks, Spoon!, Squash Racket, Steven X, Stompin' Tom, Stor stark7, Surge79uwf, Susurrus, THE KING, TJ Spyke, Tanet, Tasc, Taw, Tdadamemd, Tec15, Tempshill, Tesseran, Tewfik, The PIPE, The Rationalist, TheCoffee, TheSun, Thorsten1, Tianasez, Tim!, Tim1357, TimElessness, Timwi, Tomwalden, Tpbradbury, Treisijs, Twinxor, UberCryxic, Ugur Basak, Ulf Heinsohn, UserVOBO, VI, Van helsing, Vanished user 03, Versus22, Viking880, Wars, Webmgr, WhisperToMe, Wikist, Woohookitty, XerKibard, Yekrats, Yggdrasilsroot, Ylee, Yngvadottir, Yossiea, ZanderSchubert, Zapvet, Zictor23, 465 anonymous edits

East Germany *Source*: http://en.wikipedia.org/w/index.php?title=East_Germany *Contributors*: 159753, 172, 19TwentyTwenty, 200rabbits, 213.84.75.xxx, 52 Pickup, 888Adolf, 88Adolf, A3RO, ABlockedUser, ADS190, AKMask, Aaron Schulz, Abe Lincoln, Abrech, Acroterion, Adambiswanger1, Adambro, Adpete, Af648, AgnosticPreachersKid, Ahoerstemeier, Aigues-Mortes, Akanemoto, Aktiver Arbeiter, Alan Liefting, Alansohn, Aldy, Alensha, Alethiareg, Alex2706, Alexkin, Alexmcfire, Alssa1, Amys, AndreasJS, AndreasPraefcke, Andrejj, Andrewpmk, Andy120290, Angr, Anárion, ApolloBoy, Arathjp, Arcy, AresAndEnyo, Arsenalthefootballchampions, ArtVandelay13, Ary29, Ashot Gabrielyan, Ashtray, Asmaybe, Aspects, Avespes, Avicennasis, AxelBoldt, Axt, Badanedwa, Baldhur, Barbaar, Bark, Barry Troost, Barryob, Ben-Zin, Bender235, Benson85, Berk, Bermicourt, Bevo74, Bhludzin, Bhoy Wonder, Big Adamsky, Big Bird, Bigdaddy1981, Bigturtle, BilCat, Bill Du, Binabik80, Biruitorul, Bletch, Blinder Seher, BlueSalo, Bnull, Bob247, Bogomolov.PL, Bolegash, Bonadea, Boomshadow, Bootleg42, Bowei Huang 2, Bramlet Abercrombie, Brandmeister (old), Branislavk, Brazzy, BrendelSignature, Brendenhull, Brenont, Brion VIBBER, Brockert, Burghardts, Burkaja, Burschenschafter, Butterboy, Bwmoll3, Byrial, Bytwerk, C0lmjl, C45207, CR7, CUSENZA Mario, Cactus.man, Caden, Canadian Bobby, Captain Future, Cautious, Ceyockey, Chanheigeorge, Chochopk, Chris Roy, Chrisjwmartin, ChristianeB, Citius Altius Fortius, Citroënist, Civil Engineer III, Cl!ckpop, Cmapm, Cmskog, Cocytus, Coemgenus, Colchicum, ColdShine, Conversion script, Courcelles, Cpk1971, Craigzomack, CrimsonFury, Crystallina, Cs-wolves, Cst17, Curps, Cvieg, Cyan, Cybercobra, D-Rock, DARTH SIDIOUS 2, DJ Clayworth, DO'Neil, Dah31, Dahn, Dailycare, Daio, Dale Arnett, Danh, DanniellaWB, Davenbelle, David Levy, Davidme, Davshul, Dayubcpd, Decltype, Den fjättrade ankan, Denniss, Der Eberswalder, Der Golem, Der Sporkmeister, DerBorg, Dewritech, Dezidor, Digwuren, DirkvdM, Discospinster, Discthrower4, DisillusionedBitterAndKnackered, Dissident, Docu, Dogaroon, Domino theory, Don4of4, Donreed, DrDaveHPP, DrFrench, Dreieich, Duncharris, ESkog, EastGermanAllStar, Eclipse KOG, Edcolins, Edge Dweller, Edivorce, Edo leitner, Edward, Edwtie, Ehrenkater, EikwaR, Ekem, El C, ElinorD, Elizard16, Elmeri B. Suokirahvi, Elockid, Elykyllek, EnOreg, Enigmaman, Ennerk, Enst38, Epaphroditus Ph. M., Ericdec, Ericdn, Evb-wiki, Eveningthought, Everyking, Ewlyahoocom, Expatkiwi, Ezhiki, Ezra hasofer, Fahrenheit451, Fahrlaessig, FamilyGuy1998, Fantastic fred, Favonian, Firkin Flying Fox, Fishface, Fixman, Fkehar, Fooloomanzoo, Formeruser-81, Fourthords, Freakofnurture, Fred Bauder, FredericII, Freivolk, FreplySpang, FreshBulletTime, Friism, Fudoreaper, Fumitol, GCarty, GRuban, Gabbe, Gahread, Galoubet, Gary Joseph, Gblaz, Gcapp1959, Geo-Loge, Gesalbte, Ghettoblaster, Glen, Gnevin, Gogo Dodo, GoldRingChip, Goldsztajn, Golgofrinchian, Good Olfactory, Gr8opinionater, Graham87, Grimble2, Guppie, Gusuku, Gzornenplatz, HADRIANVS, HDCase, Hajduk, Haldraper, Halibutt, HangingCurve, HarveyWilliams, Heaviestcat, HectorRodriguez, Henrig, Herzliyya, Historian932, Hitler's Dick, Hixteilchen, Hmains, Hobartimus, Horses In The Sky, II MusLiM HyBRiD II, INkubusse, Ida Shaw, Improv, Impulsion, Infrogmation, Ingolfson, Ingsoc, Iohannes Animosus, IsarSteve, Isolani, J.delanoy, JCarriker, JETM, JNZ, JaGa, Jacob Lundberg, Jacurek, Jade Knight, JakeH07, Jakezing, Jane Herman, Jao, Jason M, Java13690, Jbs36, Jd2718, JeanJPoirier, Jeff Song, Jeremy Visser, JetLover, Jevoite, Jfruh, Jiang, Jim.henderson, Jimaginator, Jj137, Jnestorius, JoDonHo, JoSePh, Joao Xavier, Joep01, JoergenB, Johann Wolfgang, John, John K, John of Reading, JohnCD, Johnwcowan, Jonathan Stanley, Jondumitru, Jonspurling, Jonte--, Jor, JorgeGG, Joseph Solis in Australia, JoshXF, Josquius, Jovianeye, Joy, Jpatokal, Jrt989, Jtanadi, Jtgerman, Jtnelson, KGasso, Kade, Kaihsu, Kajolus, Kami888, Kanags, Kanjawe, Kar98, KarlFrei, Kay Körner, Kbrose, Kelly Martin, Ken Gallager, Kevin W., Khatru2, Khoikhoi, Kieranrichard16, Kimdime, Kimse, King of Hearts, Kingal86, Kingturtle, Kiske, Kiteinthewind, Kjetil r, Kkostagiannis, Klingon83, Knarf-bz, Knepflerle, Knotsfalcon, KnowledgeOfSelf, Koavf, Kojozone, Kosebamse, Kpjas, Krash, Kristallstadt, Krj373, Krun, Kudpung, Kurykh, Kusma, Kwamikagami, Kwekubo, Kwertii, LMB, Lacrimosus, Ladysmithrose, Lajsikonik, Langohio, Larrytheordinarydragon, Lawrencegold, Leafnode, LeeG, LeedsKing, Lfh, Li-sung, Librarymaven, Lightmouse, LilHelpa, Lipedia, Littlealien182, Lmhooo, Lokaydokay, Lord khadgar05, Lotje, LouisianaFan, Lucken, Lukemcguire, Luminella, LutzBruno, Lwalt, M Vilez, MECU, MER-C, MJCdetroit, MKLPTR, MKers06, MacRusgail, Mackensen, Madanite17, Madcynic, Magioladitis, MalcontentUK, Malljaja, Manocihr, Marcus Wendel, Marek69, Margana, Maria Stella, Martijn faassen, Martin Gazdík, Massimo Catarinella, Mattbr, Matthead, MattiPaavola, Mav, Max rspct, Mboverload, Mclay1, Menchi, Mervyn,

Metropolitan90, Mewulwe, Mhazard9, Mic, Michael Bednarek, Michael Kümmling, Michael Snow, Michaelwuzthere, Mickeylove73, Miguelitoterry, Mike1971inter, Mikko Paananen, Miroslawa, Mister Jinxy, Miyokan, Mkarau, Morton devonshire, Morven, Morwen, Mosedschurte, Mpeylo, Mrdie, Mtaylor848, MuZemike, Myanw, NSK Nikolaos S. Karastathis, Naive cynic, Nat, Nehrams2020, Neko-chan, Neolandes, Nero the second, Niceguyedc, Nickel Chromo, Nico, Nightflyer, Nikio, Nikodemos, Nima Farid, Njuuton, Nkcs, Nobody of Consequence, Norum, Nricardo, Nufy8, Numbo3, OOODDD, Octavian2009, Oknazevad, Oldbeforehistime, Olessi, Oliver202, Olivier, Oops74, Opponent, Orangerider, Orbmc101, Oreo Priest, OwenBlacker, P1nkfl0yd, PBR Recordings, PZFUN, Paddemus, Pafcool2, Pajfarmor, Palferoy, Patrick, Paul August, Paul Benjamin Austin, Paul Drye, Paul-L, Paul.h, PaulHanson, Paxse, Pearle, Pedro, Peruvianllama, Peteboro, Petri Krohn, Pfctdayelise, Phidauex, Phil1988, PhilKnight, Philip Baird Shearer, Philip Cross, Philip Trueman, Phoenix of9, Piotrus, Pischdi Hufnagel, Pokrajac, Pol098, Polylerus, ProhibitOnions, Puddhe, Quadell, Quiensabe, Qutezuce, R Lowry, R'n'B, R-41, R9tgokunks, RM21, Ramon1928, Randroide, Rapido, Raul654, Rayhou, RazorICE, Rd232, Readro, Redvers, Renata, Retired username, Rex Germanus, Ric36, Rich Farmbrough, Richiguada, RickK, Rillian, Rjensen, Rjwilmsi, Rlandmann, Rmhermen, Roadrunner, Robbie098, Robert Clausen, Robert Weemeyer, Robertb-dc, Rocastelo, RockMFR, RolandR, Romanm, Ronhjones, Roy da Vinci, Ruby Tuesday ALMWR, Ruhrjung, RuiMalheiro, Ruy Lopez, SG, SGD-LFC, SGGH, SH84, SWAdair, Saintswithin, Saltdeanmark, Sam Pointon, Sanders muc, Sandman, Sandstein, Sardanaphalus, SaxTeacher, Sc147, Sca, Sce03066, Schabowski, Schneelocke, SchuminWeb, Scipius, Scotsboyuk, SeNeKa, Search4Lancer, Secfan, Secretlondon, Senjuto, Septembermorgen, Series, Sgeureka, Shellwood, Shiftworker, ShiningEyes, Shinpah1, Shockingeye, Shogatetus, Sign language, Sillyfolkboy, Silverhorse, Simson, Skew-t, Skips, Skäpperöd, Slab stone, Sleigh, Sliggy, Smith2006, Snaildoc, SnowFire, Snowdog, Solmt, Soman, Soska, Space Cadet, Spark240, Speedboy Salesman, Spook`, Spoom, Spoonkymonkey, Spundun, StYxXx, Stagehand, StalwartUK, SteinbDJ, Stephen G. Brown, Steroid Expert, Stevenscollege, Stor stark7, Stormscape, Str1977, Strassenbahn, Summersault, Superzohar, SyL64, TDC, TFOWR, TJive, TRAJAN 117, Taboo Tongue, Tasc, Tec15, Tesscass, The Epopt, Tide rolls, Tim1357, Timc, Timneu22, Timrollpickering, Timurgabaidulin, Titoxd, Tmafarrell, Tnmusikherr, Tobias Conradi, Tom, Tommertron, Tony Papard, Tony1, Topory, Toscho, Tpbradbury, Trackinfo, Transfinite, Trevor MacInnis, Trey Stone, Trieste, Tristessa de St Ange, Triwbe, Troublemaker1973, Trusilver, Trust Is All You Need, Tsjackso, Tulandro, Turian, Tvtrojan11, Ubardak, Ugen64, Ulf Heinsohn, Ulfbastel, Ulflarsen, Unreal7, Urhixidur, Urmas, Uthlande, Valip, VampWillow, Van helsing, Vancouverguy, Vanished user 03, Vanjagenije, Varizer, Varlaam, Versus22, Vicki Rosenzweig, Vis-a-visconti, Vision Thing, VolatileChemical, WRK, Wasbeer, Wavelength, Wee Jimmy, Weerasad, Wiggy!, Wik, WikiDon, Wikist, Wilfried Derksen, Willy turner, Witt E Pseudonym, Wmahan, Wo.luren, WojPob, Woodstein52, Woohookitty, Ww2censor, XJamRastafire, Xenophon777, XerKibard, Xgmx2, Xihr, Y1997xf11, YSSYguy, Yamamoto Ichiro, Yankee Rajput, YeshuaDavid, Yoda1893, Youcandanceifyouwantto, Yyy, Zeisseng, Zenswashbuckler, Ziek163, Zocky, Zoe, Zoicon5, Zondor, Zundark, Zzuuzz, £20, °, Σ, 1000 anonymous edits

Treptower Park *Source*: http://en.wikipedia.org/w/index.php?title=Treptower_Park *Contributors*: Altenmann, Amigadave, BLueFiSH.as, Barliner, Bronks, Cmapm, CommonsDelinker, Dapoloplayer, David1955, DerBorg, Deror avi, Ernst.schnell, Freenation, Fwb44, GregorB, Impaciente, Jauhienij, Kwamikagami, Manusmanus, Mdkoch84, Meissen, Ned-kogar, Palnordseth, ProhibitOnions, Sardanaphalus, Stemonitis, TallNapoleon, Tpick1, Victorgrigas, Zozza, 27 anonymous edits

Berlin Warschauer Straße station *Source*: http://en.wikipedia.org/w/index.php?title=Berlin_Warschauer_Stra%C3%9Fe_station *Contributors*: AalanTG, Angelpeream, Angr, Anorak2, Bermicourt, Cate, DerBorg, Eco84, Gene Nygaard, GrahamHardy, Grahamec, HennessyC, Ianb, Jared Preston, Jumbo1435, Keatinga, LeonardoRob0t, MFantti, Mackensen, PZFUN, ProhibitOnions, Rd232, Schmock, Severo, Starbois, Weasel, Xyboi, 10 anonymous edits

Image Sources, Licenses and Contributors

Image:Bahnsteig Bahnhof Ostkreuz, Berlin.jpg *Source*: http://en.wikipedia.org/w/index.php?title=File:Bahnsteig_Bahnhof_Ostkreuz,_Berlin.jpg *License*: unknown *Contributors*: w:de:user:QueryzoQueryzo

Image:GleiskreuzungOstkreuzBerlin.jpg *Source*: http://en.wikipedia.org/w/index.php?title=File:GleiskreuzungOstkreuzBerlin.jpg *License*: unknown *Contributors*: User:Túrelio

File:Overview Berlin.jpg *Source*: http://en.wikipedia.org/w/index.php?title=File:Overview_Berlin.jpg *License*: unknown *Contributors*: © Berlin Partner/FTB-Werbefotografie/Myself (Montage)

File:Flag_of_Berlin.svg *Source*: http://en.wikipedia.org/w/index.php?title=File:Flag_of_Berlin.svg *License*: unknown *Contributors*: Flaggenentwurf: unbekannt diese Datei: Jwnabd

File:Coat of arms of Berlin.svg *Source*: http://en.wikipedia.org/w/index.php?title=File:Coat_of_arms_of_Berlin.svg *License*: unknown *Contributors*: Wappenentwurf: diese Datei: Jwnabd

File:Berlin in Germany and EU.png *Source*: http://en.wikipedia.org/w/index.php?title=File:Berlin_in_Germany_and_EU.png *License*: unknown *Contributors*: User:Athinaios, User:Ssolbergj, User:Unify

File:Speaker Icon.svg *Source*: http://en.wikipedia.org/w/index.php?title=File:Speaker_Icon.svg *License*: unknown *Contributors*: Blast, G.Hagedorn, Mobius, 2 anonymous edits

File:ZLB-Berliner Ansichten-Januar.jpg *Source*: http://en.wikipedia.org/w/index.php?title=File:ZLB-Berliner_Ansichten-Januar.jpg *License*: unknown *Contributors*: J. M. F. Schmidt

File:Friedrich Zweite Alt.jpg *Source*: http://en.wikipedia.org/w/index.php?title=File:Friedrich_Zweite_Alt.jpg *License*: unknown *Contributors*: Cherubino, Gryffindor, Interpretix, Maksim, 1 anonymous edits

File:Berlin Unter den Linden Victoria Hotel um 1900.jpg *Source*: http://en.wikipedia.org/w/index.php?title=File:Berlin_Unter_den_Linden_Victoria_Hotel_um_1900.jpg *License*: unknown *Contributors*: Bdk, Beek100, Finavon, Jan Arkesteijn, Janericloebe, Jcornelius, Kaganer, Kirschblut, Kurpfalzbilder.de, MB-one, Mogelzahn, Para

File:Potsdamer Platz 1945.jpg *Source*: http://en.wikipedia.org/w/index.php?title=File:Potsdamer_Platz_1945.jpg *License*: unknown *Contributors*: Michael M. Dean (Canada. Dept. of National Defence/Ministère de la défense nationale)

File:Berlinermauer.jpg *Source*: http://en.wikipedia.org/w/index.php?title=File:Berlinermauer.jpg *License*: unknown *Contributors*: Noir

File:Thefalloftheberlinwall1989.JPG *Source*: http://en.wikipedia.org/w/index.php?title=File:Thefalloftheberlinwall1989.JPG *License*: unknown *Contributors*: Unknown photographer, Reproduction by Lear 21

File:SkylineBerlin.JPG *Source*: http://en.wikipedia.org/w/index.php?title=File:SkylineBerlin.JPG *License*: unknown *Contributors*: Angr, Jonathan Groß, MB-one, Mogelzahn, R. Engelhardt, Sven Scharr, 1 anonymous edits

File:Berlin.svg *Source*: http://en.wikipedia.org/w/index.php?title=File:Berlin.svg *License*: unknown *Contributors*: Alexrk2, Leit, Puregenius, Ratatosk, Srittau, 6 anonymous edits

File:Über den Dächern von Berlin.jpg *Source*: http://en.wikipedia.org/w/index.php?title=File:Über_den_Dächern_von_Berlin.jpg *License*: unknown *Contributors*: User:Bleppo

File:Luftbild bln-schmoeckwitz.jpg *Source*: http://en.wikipedia.org/w/index.php?title=File:Luftbild_bln-schmoeckwitz.jpg *License*: unknown *Contributors*: Matthias Renner (Marathoni62)

File:Berlin night.jpg *Source*: http://en.wikipedia.org/w/index.php?title=File:Berlin_night.jpg *License*: unknown *Contributors*: Robert Debowski

File:Neue nationalgalerie treppen.jpg *Source*: http://en.wikipedia.org/w/index.php?title=File:Neue_nationalgalerie_treppen.jpg *License*: unknown *Contributors*: Mangan2002 (sv.wikipedia.org)

File:Brandenburger Tor Nachts.JPG *Source*: http://en.wikipedia.org/w/index.php?title=File:Brandenburger_Tor_Nachts.JPG *License*: unknown *Contributors*: User:Knipptang

File:Berlin potsdamer platz db haus atnight.jpg *Source*: http://en.wikipedia.org/w/index.php?title=File:Berlin_potsdamer_platz_db_haus_atnight.jpg *License*: unknown *Contributors*: janine pohl

File:Gedächtniskirche1.JPG *Source*: http://en.wikipedia.org/w/index.php?title=File:Gedächtniskirche1.JPG *License*: unknown *Contributors*: BLueFiSH.as, PetrusSilesius, Wst

File:Schloss Charlottenburg 2005 282.JPG *Source*: http://en.wikipedia.org/w/index.php?title=File:Schloss_Charlottenburg_2005_282.JPG *License*: unknown *Contributors*: User:Noebu

File:Reichstag building Berlin view from west before sunset.jpg *Source*: http://en.wikipedia.org/w/index.php?title=File:Reichstag_building_Berlin_view_from_west_before_sunset.jpg *License*: unknown *Contributors*: User:JürgenMatern

File:Wowereit2008.JPG *Source*: http://en.wikipedia.org/w/index.php?title=File:Wowereit2008.JPG *License*: unknown *Contributors*: User:KJohansson

File:BerlinDistricts.svg *Source*: http://en.wikipedia.org/w/index.php?title=File:BerlinDistricts.svg *License*: unknown *Contributors*: User:Angr

File:Flag of the United States.svg *Source*: http://en.wikipedia.org/w/index.php?title=File:Flag_of_the_United_States.svg *License*: unknown *Contributors*: Anomie

File:Flag of Hungary.svg *Source*: http://en.wikipedia.org/w/index.php?title=File:Flag_of_Hungary.svg *License*: unknown *Contributors*: User:SKopp

File:Flag of Japan.svg *Source*: http://en.wikipedia.org/w/index.php?title=File:Flag_of_Japan.svg *License*: unknown *Contributors*: Anomie

File:Flag of France.svg *Source*: http://en.wikipedia.org/w/index.php?title=File:Flag_of_France.svg *License*: unknown *Contributors*: Anomie

File:Flag of Belgium (civil).svg *Source*: http://en.wikipedia.org/w/index.php?title=File:Flag_of_Belgium_(civil).svg *License*: unknown *Contributors*: Bean49, David Descamps, Dbenbenn, Denelson83, Evanc0912, Fry1989, Gabriel trzy, Howcome, Ms2ger, Nightstallion, Oreo Priest, Rocket000, Rodejong, Sir Iain, ThomasPusch, Warddr, Zscout370, 4 anonymous edits

File:Flag of Argentina.svg *Source*: http://en.wikipedia.org/w/index.php?title=File:Flag_of_Argentina.svg *License*: unknown *Contributors*: User:Dbenbenn

File:Flag of Spain.svg *Source*: http://en.wikipedia.org/w/index.php?title=File:Flag_of_Spain.svg *License*: unknown *Contributors*: Anomie

File:Flag of Indonesia.svg *Source*: http://en.wikipedia.org/w/index.php?title=File:Flag_of_Indonesia.svg *License*: unknown *Contributors*: User:Gabbe, User:SKopp

File:Flag of the Czech Republic.svg *Source*: http://en.wikipedia.org/w/index.php?title=File:Flag_of_the_Czech_Republic.svg *License*: unknown *Contributors*: special commission (of code): SVG version by cs:-xfi-. Colors according to Appendix No. 3 of czech legal Act 3/1993. cs:Zirland.

File:Flag of Turkey.svg *Source*: http://en.wikipedia.org/w/index.php?title=File:Flag_of_Turkey.svg *License*: unknown *Contributors*: User:Dbenbenn

File:Flag of Uzbekistan.svg *Source*: http://en.wikipedia.org/w/index.php?title=File:Flag_of_Uzbekistan.svg *License*: unknown *Contributors*: User:Zscout370

File:Flag of Namibia.svg *Source*: http://en.wikipedia.org/w/index.php?title=File:Flag_of_Namibia.svg *License*: unknown *Contributors*: User:Vzb83

File:Flag of Poland.svg *Source*: http://en.wikipedia.org/w/index.php?title=File:Flag_of_Poland.svg *License*: unknown *Contributors*: Anomie, Mifter

File:Flag of Mexico.svg *Source*: http://en.wikipedia.org/w/index.php?title=File:Flag_of_Mexico.svg *License*: unknown *Contributors*: User:AlexCovarrubias

File:Flag of the United Kingdom.svg *Source*: http://en.wikipedia.org/w/index.php?title=File:Flag_of_the_United_Kingdom.svg *License*: unknown *Contributors*: Anomie, Mifter

File:Flag of Russia.svg *Source*: http://en.wikipedia.org/w/index.php?title=File:Flag_of_Russia.svg *License*: unknown *Contributors*: Anomie

File:Flag of the People's Republic of China.svg *Source*: http://en.wikipedia.org/w/index.php?title=File:Flag_of_the_People's_Republic_of_China.svg *License*: unknown *Contributors*: User:Denelson83, User:SKopp, User:Shizhao, User:Zscout370

File:ICC Funkturm Berlin.jpg *Source*: http://en.wikipedia.org/w/index.php?title=File:ICC_Funkturm_Berlin.jpg *License*: unknown *Contributors*: User:Bleppo

File:Flag of Berlin.svg *Source*: http://en.wikipedia.org/w/index.php?title=File:Flag_of_Berlin.svg *License*: unknown *Contributors*: Flaggenentwurf: unbekannt diese Datei: Jwnabd

File:Flag of Germany.svg *Source*: http://en.wikipedia.org/w/index.php?title=File:Flag_of_Germany.svg *License*: unknown *Contributors*: Anomie

File:Flag of Europe.svg *Source*: http://en.wikipedia.org/w/index.php?title=File:Flag_of_Europe.svg *License*: unknown *Contributors*: User:-xfi-, User:Dbenbenn, User:Funakoshi, User:Jeltz, User:Nightstallion, User:Paddu, User:Verdy p, User:Zscout370

File:Air Berlin B737 D-ABAA.jpg *Source*: http://en.wikipedia.org/w/index.php?title=File:Air_Berlin_B737_D-ABAA.jpg *License*: unknown *Contributors*: Axel.Mauruszat, Gomera-b, JuergenL, MB-one, WhisperToMe

File:BlnHauptbahnhof28.jpg *Source*: http://en.wikipedia.org/w/index.php?title=File:BlnHauptbahnhof28.jpg *License*: unknown *Contributors*: User:Angr

Image:EDDT Empfang & Tower.jpg *Source*: http://en.wikipedia.org/w/index.php?title=File:EDDT_Empfang_&_Tower.jpg *License*: unknown *Contributors*: User:MB-one

Image:BBI 2010-07-23 5.JPG *Source*: http://en.wikipedia.org/w/index.php?title=File:BBI_2010-07-23_5.JPG *License*: unknown *Contributors*: Michael F. Mehnert

File:Berlin HKW Mitte.jpg *Source*: http://en.wikipedia.org/w/index.php?title=File:Berlin_HKW_Mitte.jpg *License*: unknown *Contributors*: Axel.Mauruszat, BLueFiSH.as, Majorly, Unify

File:Charite 1a.jpg *Source*: http://en.wikipedia.org/w/index.php?title=File:Charite_1a.jpg *License*: unknown *Contributors*: MB-one, Norro, 1 anonymous edits

File:Berlin Population Development 1880-2007.svg *Source*: http://en.wikipedia.org/w/index.php?title=File:Berlin_Population_Development_1880-2007.svg *License*: unknown *Contributors*: user:Unify

File:Berliner Dom Spreeinsel.jpg *Source*: http://en.wikipedia.org/w/index.php?title=File:Berliner_Dom_Spreeinsel.jpg *License*: unknown *Contributors*: User:Dnsob

File:Humboldt monument.jpg *Source*: http://en.wikipedia.org/w/index.php?title=File:Humboldt_monument.jpg *License*: unknown *Contributors*: Rudi Heim from Hannover, Germany

File:Solartankstelle.jpg *Source*: http://en.wikipedia.org/w/index.php?title=File:Solartankstelle.jpg *License*: unknown *Contributors*: Busso v. Bismarck

File:Bodemuseum - Front.jpeg *Source*: http://en.wikipedia.org/w/index.php?title=File:Bodemuseum_-_Front.jpeg *License*: unknown *Contributors*: http://flickr.com/photos/dalbera/

File:Berlin - Axel-Springer-Hochhaus.jpg *Source*: http://en.wikipedia.org/w/index.php?title=File:Berlin_-_Axel-Springer-Hochhaus.jpg *License*: unknown *Contributors*: Johann H. Addicks

File:Cate blanchett.jpg *Source*: http://en.wikipedia.org/w/index.php?title=File:Cate_blanchett.jpg *License*: unknown *Contributors*: User:Sashandre

File:Buddy Bear One-World Bär.JPG *Source*: http://en.wikipedia.org/w/index.php?title=File:Buddy_Bear_One-World_Bär.JPG *License*: unknown *Contributors*: Floridi

File:In Berghain.jpg *Source*: http://en.wikipedia.org/w/index.php?title=File:In_Berghain.jpg *License*: unknown *Contributors*: User:Marcheur1976

File:Erster Spreestrand.JPG *Source*: http://en.wikipedia.org/w/index.php?title=File:Erster_Spreestrand.JPG *License*: unknown *Contributors*: selbst

File:Jewishmuseumberlin2007.jpg *Source*: http://en.wikipedia.org/w/index.php?title=File:Jewishmuseumberlin2007.jpg *License*: unknown *Contributors*: Original uploader was Lear 21 at en.wikipedia. Later version(s) were uploaded by Angr at en.wikipedia.

File:Ishtar Gate at Berlin Museum.jpg *Source*: http://en.wikipedia.org/w/index.php?title=File:Ishtar_Gate_at_Berlin_Museum.jpg *License*: unknown *Contributors*: photo by Rictor Norton

File:Rattle BPH-Rittershaus1-Wikipedia.jpg *Source*: http://en.wikipedia.org/w/index.php?title=File:Rattle_BPH-Rittershaus1-Wikipedia.jpg *License*: unknown *Contributors*: Monika Rittershaus / per OTRS

File:Haus der Kulturen der Welt Nachtaufnahme.jpg *Source*: http://en.wikipedia.org/w/index.php?title=File:Haus_der_Kulturen_der_Welt_Nachtaufnahme.jpg *License*: unknown *Contributors*: holger doelle

File:CurryWurst.jpg *Source*: http://en.wikipedia.org/w/index.php?title=File:CurryWurst.jpg *License*: unknown *Contributors*: User:Agerio

File:Giraffe-berlin-zoo.jpg *Source*: http://en.wikipedia.org/w/index.php?title=File:Giraffe-berlin-zoo.jpg *License*: unknown *Contributors*: Adamantios

File:Berlin Olympiastadion during footballmatch herta bsc berlin vs borussia dortmund 02 20070421.jpg *Source*: http://en.wikipedia.org/w/index.php?title=File:Berlin_Olympiastadion_during_footballmatch_herta_bsc_berlin_vs_borussia_dortmund_02_20070421.jpg *License*: unknown *Contributors*: -

File:Berlin marathon.jpg *Source*: http://en.wikipedia.org/w/index.php?title=File:Berlin_marathon.jpg *License*: unknown *Contributors*: User:KJohansson

File:Marlene Dietrich in Stage Fright trailer.jpg *Source*: http://en.wikipedia.org/w/index.php?title=File:Marlene_Dietrich_in_Stage_Fright_trailer.jpg *License*: unknown *Contributors*: Trailer screenshot

File:S-Bahn-Logo.svg *Source*: http://en.wikipedia.org/w/index.php?title=File:S-Bahn-Logo.svg *License*: unknown *Contributors*: User:LeCornichon

File:Train station Berlin Friedrichstrasse 5.jpg *Source*: http://en.wikipedia.org/w/index.php?title=File:Train_station_Berlin_Friedrichstrasse_5.jpg *License*: unknown *Contributors*: Akinom, BLueFiSH.as, Beek100, FlickreviewR, Jcornelius, Sebastian Wallroth

Image:Topographischer Netzplan der S-Bahn Berlin.png *Source*: http://en.wikipedia.org/w/index.php?title=File:Topographischer_Netzplan_der_S-Bahn_Berlin.png *License*: unknown *Contributors*: Maximilian Dörrbecker (Chumwa)

File:Berlin S1.svg *Source*: http://en.wikipedia.org/w/index.php?title=File:Berlin_S1.svg *License*: unknown *Contributors*: User:NordNordWest

File:Berlin S2.svg *Source*: http://en.wikipedia.org/w/index.php?title=File:Berlin_S2.svg *License*: unknown *Contributors*: User:NordNordWest

File:Berlin S25.svg *Source*: http://en.wikipedia.org/w/index.php?title=File:Berlin_S25.svg *License*: unknown *Contributors*: User:NordNordWest

File:Berlin S3.svg *Source*: http://en.wikipedia.org/w/index.php?title=File:Berlin_S3.svg *License*: unknown *Contributors*: User:NordNordWest

File:Berlin S41.svg *Source*: http://en.wikipedia.org/w/index.php?title=File:Berlin_S41.svg *License*: unknown *Contributors*: User:NordNordWest

File:Berlin S42.svg *Source*: http://en.wikipedia.org/w/index.php?title=File:Berlin_S42.svg *License*: unknown *Contributors*: User:NordNordWest

File:Berlin S45.svg *Source*: http://en.wikipedia.org/w/index.php?title=File:Berlin_S45.svg *License*: unknown *Contributors*: User:NordNordWest

File:Berlin S46.svg *Source*: http://en.wikipedia.org/w/index.php?title=File:Berlin_S46.svg *License*: unknown *Contributors*: User:NordNordWest

File:Berlin S47.svg *Source*: http://en.wikipedia.org/w/index.php?title=File:Berlin_S47.svg *License*: unknown *Contributors*: User:NordNordWest

File:Berlin S5.svg *Source*: http://en.wikipedia.org/w/index.php?title=File:Berlin_S5.svg *License*: unknown *Contributors*: User:NordNordWest

File:Berlin S7.svg *Source*: http://en.wikipedia.org/w/index.php?title=File:Berlin_S7.svg *License*: unknown *Contributors*: User:NordNordWest

File:Berlin S75.svg *Source*: http://en.wikipedia.org/w/index.php?title=File:Berlin_S75.svg *License*: unknown *Contributors*: User:NordNordWest

File:Berlin S8.svg *Source*: http://en.wikipedia.org/w/index.php?title=File:Berlin_S8.svg *License*: unknown *Contributors*: User:NordNordWest

File:Berlin S85.svg *Source*: http://en.wikipedia.org/w/index.php?title=File:Berlin_S85.svg *License*: unknown *Contributors*: User:NordNordWest

File:Berlin S9.svg *Source*: http://en.wikipedia.org/w/index.php?title=File:Berlin_S9.svg *License*: unknown *Contributors*: User:NordNordWest

Image:S-Bahn Berlin Baureihe 477.jpg *Source*: http://en.wikipedia.org/w/index.php?title=File:S-Bahn_Berlin_Baureihe_477.jpg *License*: unknown *Contributors*: Michael Dittrich

Image:Train station Berlin Alexanderplatz 1999 pixelquelle.jpg *Source*: http://en.wikipedia.org/w/index.php?title=File:Train_station_Berlin_Alexanderplatz_1999_pixelquelle.jpg *License*: unknown *Contributors*: Hans-Georg Weimar from pixelquelle.de

File:Bf-b-anhalterbf.jpg *Source*: http://en.wikipedia.org/w/index.php?title=File:Bf-b-anhalterbf.jpg *License*: unknown *Contributors*: A.Savin

Image:S-Bahn Berlin Baureihe 481.jpg *Source*: http://en.wikipedia.org/w/index.php?title=File:S-Bahn_Berlin_Baureihe_481.jpg *License*: unknown *Contributors*: Jorges

File:LIRR sampler electric and diesel services.jpg *Source*: http://en.wikipedia.org/w/index.php?title=File:LIRR_sampler_electric_and_diesel_services.jpg *License*: unknown *Contributors*: User:AEMoreira042281

File:İZBAN Train.JPG *Source*: http://en.wikipedia.org/w/index.php?title=File:İZBAN_Train.JPG *License*: unknown *Contributors*: Denizcan Dorsam. Original uploader was Central Data Bank at en.wikipedia

File:British Rail Class 318 264 r.jpg *Source*: http://en.wikipedia.org/w/index.php?title=File:British_Rail_Class_318_264_r.jpg *License*: unknown *Contributors*: User:Dave souza

File:Renfe Cercanías New Train.jpg *Source*: http://en.wikipedia.org/w/index.php?title=File:Renfe_Cercanías_New_Train.jpg *License*: unknown *Contributors*: Aloriel, Balbo, Loqu, Yrithinnd

File:Unit 376029 at Slade Green.JPG *Source*: http://en.wikipedia.org/w/index.php?title=File:Unit_376029_at_Slade_Green.JPG *License*: unknown *Contributors*: User:Sunil060902

File:Korail CommutorDieselCar.jpg *Source*: http://en.wikipedia.org/w/index.php?title=File:Korail_CommutorDieselCar.jpg *License*: unknown *Contributors*: G43, LERK, Peremen

File:423-Vollzug Main-Weser-Bahn.jpg *Source*: http://en.wikipedia.org/w/index.php?title=File:423-Vollzug_Main-Weser-Bahn.jpg *License*: unknown *Contributors*: MdE, Melkom

File:GO Train at Exhibition GO Station with a view of CN Tower.jpg *Source*: http://en.wikipedia.org/w/index.php?title=File:GO_Train_at_Exhibition_GO_Station_with_a_view_of_CN_Tower.jpg *License*: unknown *Contributors*: Original uploader was SFernandez at en.wikipedia

File:ЭД4МК-022.JPG *Source*: http://en.wikipedia.org/w/index.php?title=File:ЭД4МК-022.JPG *License*: unknown *Contributors*: User:Max071086

Image:Mumbai Train.JPG *Source*: http://en.wikipedia.org/w/index.php?title=File:Mumbai_Train.JPG *License*: unknown *Contributors*: Original uploader was Nikkul at en.wikipedia

File:Cityrail-millennium-M32-ext.jpg *Source*: http://en.wikipedia.org/w/index.php?title=File:Cityrail-millennium-M32-ext.jpg *License*: unknown *Contributors*: User:Alexanderino

File:Class 83 KTM Komuter train, Kuala Lumpur.jpg *Source*: http://en.wikipedia.org/w/index.php?title=File:Class_83_KTM_Komuter_train,_Kuala_Lumpur.jpg *License*: unknown *Contributors*: User:Two hundred percent

File:Lodi TSR S1.JPG *Source*: http://en.wikipedia.org/w/index.php?title=File:Lodi_TSR_S1.JPG *License*: unknown *Contributors*: User:Friedrichstrasse

File:Metra City of Woodstock in Deerfield.jpg *Source*: http://en.wikipedia.org/w/index.php?title=File:Metra_City_of_Woodstock_in_Deerfield.jpg *License*: unknown *Contributors*: Keeleysam at en.wikipedia

File:JRE-E231-500-for-JRyamanote-line.jpg *Source*: http://en.wikipedia.org/w/index.php?title=File:JRE-E231-500-for-JRyamanote-line.jpg *License*: unknown *Contributors*: Hohoho, LERK,

File:KRL Jabotabek 7117 Gambir.jpg *Source*: http://en.wikipedia.org/w/index.php?title=File:KRL_Jabotabek_7117_Gambir.jpg *License*: unknown *Contributors*: User:Sushiya

Image:Ny pendeltag stockholm.jpg *Source*: http://en.wikipedia.org/w/index.php?title=File:Ny_pendeltag_stockholm.jpg *License*: unknown *Contributors*: User:Nordelch

File:Z20500lm.jpg *Source*: http://en.wikipedia.org/w/index.php?title=File:Z20500lm.jpg *License*: unknown *Contributors*: User:Romain9247

File:Toshiba Electrico Ferrocarril Sarmiento Haedo.JPG *Source*: http://en.wikipedia.org/w/index.php?title=File:Toshiba_Electrico_Ferrocarril_Sarmiento_Haedo.JPG *License*: unknown *Contributors*: User:Albasmalko

File:Flag of Germany (1946-1949).svg *Source*: http://en.wikipedia.org/w/index.php?title=File:Flag_of_Germany_(1946-1949).svg *License*: unknown *Contributors*: Alkari, B1mbo, Ejdzej, Fry1989, Jwnabd, Mattes, Phlegmatic, Slomox, Ukilot, WerWil, 3 anonymous edits

File:Flag of East Berlin (1956-1990).svg *Source*: http://en.wikipedia.org/w/index.php?title=File:Flag_of_East_Berlin_(1956-1990).svg *License*: unknown *Contributors*: User:Madden

File:Coat of arms of Berlin (1935).svg *Source*: http://en.wikipedia.org/w/index.php?title=File:Coat_of_arms_of_Berlin_(1935).svg *License*: unknown *Contributors*: User:Madden

File:Occupied Berlin.svg *Source*: http://en.wikipedia.org/w/index.php?title=File:Occupied_Berlin.svg *License*: unknown *Contributors*: FDV, Fry1989, Langläufer, Leit, Petr Dlouhý, Stefan-Xp

File:Marx-engels-platz.jpg *Source*: http://en.wikipedia.org/w/index.php?title=File:Marx-engels-platz.jpg *License*: unknown *Contributors*: Arne List

File:EastBerlinBoroughs.png *Source*: http://en.wikipedia.org/w/index.php?title=File:EastBerlinBoroughs.png *License*: unknown *Contributors*: Alexrk2, Alfanje, Complex01, Edward, Werckmeister

File:Karl-Marx-Allee Block C Nord Berlin April 2006 060.jpg *Source*: http://en.wikipedia.org/w/index.php?title=File:Karl-Marx-Allee_Block_C_Nord_Berlin_April_2006_060.jpg *License*: unknown *Contributors*: User:Gryffindor

File:Lenindecoration.JPG *Source*: http://en.wikipedia.org/w/index.php?title=File:Lenindecoration.JPG *License*: unknown *Contributors*: Adam Carr at en.wikipedia

File:Veggdekorasjon.JPG *Source*: http://en.wikipedia.org/w/index.php?title=File:Veggdekorasjon.JPG *License*: unknown *Contributors*: Adam Carr at en.wikipedia

File:Soviet War Memorial Park.JPG *Source*: http://en.wikipedia.org/w/index.php?title=File:Soviet_War_Memorial_Park.JPG *License*: unknown *Contributors*: User:Adam Carr

File:Cafe Moskau.JPG *Source*: http://en.wikipedia.org/w/index.php?title=File:Cafe_Moskau.JPG *License*: unknown *Contributors*: Achim Raschka, AndreasPraefcke, Gryffindor, Matt314, PDD, 1 anonymous edits

File:Dismantling of the Palace of the Republic.JPG *Source*: http://en.wikipedia.org/w/index.php?title=File:Dismantling_of_the_Palace_of_the_Republic.JPG *License*: unknown *Contributors*: User:Adam Carr

File:New Synagogue, East Berlin.JPG *Source*: http://en.wikipedia.org/w/index.php?title=File:New_Synagogue,_East_Berlin.JPG *License*: unknown *Contributors*: User:Adam Carr

File:Weberwiese Berlin April 2006 147.jpg *Source*: http://en.wikipedia.org/w/index.php?title=File:Weberwiese_Berlin_April_2006_147.jpg *License*: unknown *Contributors*: User:Gryffindor

File:VolksbühBerlJan08.JPG *Source*: http://en.wikipedia.org/w/index.php?title=File:VolksbühBerlJan08.JPG *License*: unknown *Contributors*: User:Alex1011

File:Berlin_Wilhelmstrasse.jpg *Source*: http://en.wikipedia.org/w/index.php?title=File:Berlin_Wilhelmstrasse.jpg *License*: unknown *Contributors*: User:BetacommandBot, User:ChristianBier, User:Gryffindor, User:Kirschblut

File:Strausberger Platz Berlin April 2006 117.jpg *Source*: http://en.wikipedia.org/w/index.php?title=File:Strausberger_Platz_Berlin_April_2006_117.jpg *License*: unknown *Contributors*: User:Gryffindor

File:Proletarian hero, Alexanderplatz june 2006.JPG *Source*: http://en.wikipedia.org/w/index.php?title=File:Proletarian_hero,_Alexanderplatz_june_2006.JPG *License*: unknown *Contributors*: Adam Carr at en.wikipedia

File:Statues of Karl Marx and Friedrich Engels.JPG *Source*: http://en.wikipedia.org/w/index.php?title=File:Statues_of_Karl_Marx_and_Friedrich_Engels.JPG *License*: unknown *Contributors*: User:Adam Carr

Image:Oberbaumbrücke mit U-Bahn.jpg *Source*: http://en.wikipedia.org/w/index.php?title=File:Oberbaumbrücke_mit_U-Bahn.jpg *License*: unknown *Contributors*: Sarah Jane at Flickr

File:Coat of arms de-be friedrichshain 1991.png *Source*: http://en.wikipedia.org/w/index.php?title=File:Coat_of_arms_de-be_friedrichshain_1991.png *License*: unknown *Contributors*: Magul, PDD, Srittau, Ulf Heinsohn

file:Germany location map.svg *Source*: http://en.wikipedia.org/w/index.php?title=File:Germany_location_map.svg *License*: unknown *Contributors*: User:NordNordWest

File:Red_pog.svg *Source*: http://en.wikipedia.org/w/index.php?title=File:Red_pog.svg *License*: unknown *Contributors*: Anomie

Image:Berlin karlmarxallee kl.jpg *Source*: http://en.wikipedia.org/w/index.php?title=File:Berlin_karlmarxallee_kl.jpg *License*: unknown *Contributors*: Achim Raschka, FRZ, Jcornelius, Ronaldino

Image:Simon-Dach-Straße Berlin.jpg *Source*: http://en.wikipedia.org/w/index.php?title=File:Simon-Dach-Straße_Berlin.jpg *License*: unknown *Contributors*: Gryffindor, Jcornelius

Image:SimonDachStr.jpg *Source*: http://en.wikipedia.org/w/index.php?title=File:SimonDachStr.jpg *License*: unknown *Contributors*: Original uploader was ProhibitOnions at en.wikipedia

Image:Jugendstil Simon-Dach-Straße Berlin 2.jpg *Source*: http://en.wikipedia.org/w/index.php?title=File:Jugendstil_Simon-Dach-Straße_Berlin_2.jpg *License*: unknown *Contributors*: Gryffindor, Jcornelius

Image:Knorrpromenade Friedrichshain Berlin.jpg *Source*: http://en.wikipedia.org/w/index.php?title=File:Knorrpromenade_Friedrichshain_Berlin.jpg *License*: unknown *Contributors*: Beek100, Gryffindor, Jcornelius

Image:Kino Intimes.jpg *Source*: http://en.wikipedia.org/w/index.php?title=File:Kino_Intimes.jpg *License*: unknown *Contributors*: User:Nicor

Image:BoxhagenerPlatzFleamarket.jpg *Source*: http://en.wikipedia.org/w/index.php?title=File:BoxhagenerPlatzFleamarket.jpg *License*: unknown *Contributors*: Original uploader was ProhibitOnions at en.wikipedia

Image:Frankfurter_Allee_2006.jpg *Source*: http://en.wikipedia.org/w/index.php?title=File:Frankfurter_Allee_2006.jpg *License*: unknown *Contributors*: Original uploader was ProhibitOnions at en.wikipedia

Image:NorthTowerFrankTor.jpg *Source*: http://en.wikipedia.org/w/index.php?title=File:NorthTowerFrankTor.jpg *License*: unknown *Contributors*: Original uploader was ProhibitOnions at en.wikipedia

Image:FrankfurterAlleefacade.jpg *Source*: http://en.wikipedia.org/w/index.php?title=File:FrankfurterAlleefacade.jpg *License*: unknown *Contributors*: Original uploader was ProhibitOnions at en.wikipedia

Image:WEBERWIESEHAUS.JPG *Source*: http://en.wikipedia.org/w/index.php?title=File:WEBERWIESEHAUS.JPG *License*: unknown *Contributors*: Original uploader was ProhibitOnions at en.wikipedia

Image:MoleculeMen BerlinerOsthafen.jpg *Source*: http://en.wikipedia.org/w/index.php?title=File:MoleculeMen_BerlinerOsthafen.jpg *License*: unknown *Contributors*: User:Túrelio

Image:Berlin-friedrichshain osthafen 20050922 437.jpg *Source*: http://en.wikipedia.org/w/index.php?title=File:Berlin-friedrichshain_osthafen_20050922_437.jpg *License*: unknown *Contributors*: User:Georg Slickers

Image:East_Side_Gallery.JPG *Source*: http://en.wikipedia.org/w/index.php?title=File:East_Side_Gallery.JPG *License*: unknown *Contributors*: Kjetil r, Mutter Erde

Image:KreuzigerSquat.jpg *Source*: http://en.wikipedia.org/w/index.php?title=File:KreuzigerSquat.jpg *License*: unknown *Contributors*: Original uploader was ProhibitOnions at en.wikipedia

Image:Coat of arms of borough Friedrichshain-Kreuzberg.svg *Source*: http://en.wikipedia.org/w/index.php?title=File:Coat_of_arms_of_borough_Friedrichshain-Kreuzberg.svg *License*: unknown *Contributors*: Wappenentwurf: unbekannt diese Datei: Jwnabd

Image:Berlin friedrichshain-kreuzberg.png *Source*: http://en.wikipedia.org/w/index.php?title=File:Berlin_friedrichshain-kreuzberg.png *License*: unknown *Contributors*: Puregenius, Srittau

File:Berlin Friedrichshain-Kreuzberg.png *Source*: http://en.wikipedia.org/w/index.php?title=File:Berlin_Friedrichshain-Kreuzberg.png *License*: unknown *Contributors*: User:BishkekRocks, User:BishkekRocks

File:Allocation of seats in the borough council of Friedrichshain-Kreuzberg (DE-2011-10-27).svg *Source*: http://en.wikipedia.org/w/index.php?title=File:Allocation_of_seats_in_the_borough_council_of_Friedrichshain-Kreuzberg_(DE-2011-10-27).svg *License*: unknown *Contributors*: Jwnabd

File:Flag of Nicaragua.svg *Source*: http://en.wikipedia.org/w/index.php?title=File:Flag_of_Nicaragua.svg *License*: unknown *Contributors*: User:Nightstallion

File:Flag of Israel.svg *Source*: http://en.wikipedia.org/w/index.php?title=File:Flag_of_Israel.svg *License*: unknown *Contributors*: AnonMoos, Bastique, Bobika, Brown spite, Captain Zizi, Cerveaugenie, Drork, Etams, Fred J, Fry1989, Geagea, Himasaram, Homo lupus, Humus sapiens, Klemen Kocjancic, Kookaburra, Luispihormiguero, Madden, Neq00, NielsF, Nightstallion, Oren neu dag, Patstuart, PeeJay2K3, Pumbaa80, Ramiy, Reisio, Rodejong, SKopp, Sceptic, SomeDudeWithAUserName, Technion, Typhix, Valentinian, Yellow up, Zscout370, 31 anonymous edits

File:Flag of Bulgaria.svg *Source*: http://en.wikipedia.org/w/index.php?title=File:Flag_of_Bulgaria.svg *License*: unknown *Contributors*: User:SKopp

Image:Berlin Lichtenberg Victoriastadt Tuchollaplatz.jpg *Source*: http://en.wikipedia.org/w/index.php?title=File:Berlin_Lichtenberg_Victoriastadt_Tuchollaplatz.jpg *License*: unknown *Contributors*: Jcornelius, Lotse

File:Berlin-Noeldnerplatz.JPG *Source*: http://en.wikipedia.org/w/index.php?title=File:Berlin-Noeldnerplatz.JPG *License*: unknown *Contributors*: User:Global Fish

File:Berlin, Rummelsburger See, Wohnanlage Hauptstrasse.jpg *Source*: http://en.wikipedia.org/w/index.php?title=File:Berlin,_Rummelsburger_See,_Wohnanlage_Hauptstrasse.jpg *License*: unknown *Contributors*: User:Beek100

File:Muensterlandplatz 2.jpg *Source*: http://en.wikipedia.org/w/index.php?title=File:Muensterlandplatz_2.jpg *License*: unknown *Contributors*: User:44penguins

File:Berlin-rummelsburg kraftwerk-klingenberg 20051017 092.jpg *Source*: http://en.wikipedia.org/w/index.php?title=File:Berlin-rummelsburg_kraftwerk-klingenberg_20051017_092.jpg *License*: unknown *Contributors*: User:Georg Slickers

file:Db-schild.svg *Source*: http://en.wikipedia.org/w/index.php?title=File:Db-schild.svg *License*: unknown *Contributors*: See below.

file:S-Bahn_Berlin_Westkreuz_Entrance.jpg *Source*: http://en.wikipedia.org/w/index.php?title=File:S-Bahn_Berlin_Westkreuz_Entrance.jpg *License*: unknown *Contributors*: Jcornelius, MB-one

Image:EastboundS7WestKreuz.jpg *Source*: http://en.wikipedia.org/w/index.php?title=File:EastboundS7WestKreuz.jpg *License*: unknown *Contributors*: User:Jef Kazak

File:Karte berlin stadtbahn.png *Source*: http://en.wikipedia.org/w/index.php?title=File:Karte_berlin_stadtbahn.png *License*: unknown *Contributors*: Chumwa, Citypeek, Dinamik, Jed, Mogelzahn, Pequod76, Sansculotte, 1 anonymous edits

File:BSicon_STR.svg *Source*: http://en.wikipedia.org/w/index.php?title=File:BSicon_STR.svg *License*: unknown *Contributors*: &

File:BSicon_SBHF.svg *Source*: http://en.wikipedia.org/w/index.php?title=File:BSicon_SBHF.svg *License*: unknown *Contributors*: $traight-$hoota, Axpde, Bernina

File:BSicon_BHF.svg *Source*: http://en.wikipedia.org/w/index.php?title=File:BSicon_BHF.svg *License*: unknown *Contributors*: Bernina & axpde

File:BSicon_SHST.svg *Source*: http://en.wikipedia.org/w/index.php?title=File:BSicon_SHST.svg *License*: unknown *Contributors*: $traight-$hoota, Axpde, Dealerofsalvation

File:Berlin U8.svg *Source*: http://en.wikipedia.org/w/index.php?title=File:Berlin_U8.svg *License*: unknown *Contributors*: User:NordNordWest

File:BSicon_HST.svg *Source*: http://en.wikipedia.org/w/index.php?title=File:BSicon_HST.svg *License*: unknown *Contributors*: $traight-$hoota, AndreyA, Axpde, Bernina

File:Berlin U2.svg *Source*: http://en.wikipedia.org/w/index.php?title=File:Berlin_U2.svg *License*: unknown *Contributors*: User:NordNordWest

File:Berlin U5.svg *Source*: http://en.wikipedia.org/w/index.php?title=File:Berlin_U5.svg *License*: unknown *Contributors*: User:NordNordWest

File:BSicon_TSBHFt.svg *Source*: http://en.wikipedia.org/w/index.php?title=File:BSicon_TSBHFt.svg *License*: unknown *Contributors*: AndreyA, Axpde, Erik Baas

File:BSicon_TBHFt.svg *Source*: http://en.wikipedia.org/w/index.php?title=File:BSicon_TBHFt.svg *License*: unknown *Contributors*: Axpde, Erik Baas

File:Berlin U6.svg *Source*: http://en.wikipedia.org/w/index.php?title=File:Berlin_U6.svg *License*: unknown *Contributors*: User:NordNordWest

File:BSicon_TSHSTt.svg *Source*: http://en.wikipedia.org/w/index.php?title=File:BSicon_TSHSTt.svg *License*: unknown *Contributors*: AndreyA, Axpde, Erik Baas

File:Berlin U55.svg *Source*: http://en.wikipedia.org/w/index.php?title=File:Berlin_U55.svg *License*: unknown *Contributors*: User:NordNordWest

File:BSicon_ÜST.svg *Source*: http://en.wikipedia.org/w/index.php?title=File:BSicon_ÜST.svg *License*: unknown *Contributors*: user:axpde

File:Berlin U9.svg *Source*: http://en.wikipedia.org/w/index.php?title=File:Berlin_U9.svg *License*: unknown *Contributors*: User:NordNordWest

File:Berlin U7.svg *Source*: http://en.wikipedia.org/w/index.php?title=File:Berlin_U7.svg *License*: unknown *Contributors*: User:NordNordWest

File:BSicon_ABZdlf.svg *Source*: http://en.wikipedia.org/w/index.php?title=File:BSicon_ABZdlf.svg *License*: unknown *Contributors*: User:Lantus

File:BSicon_KRZo.svg *Source*: http://en.wikipedia.org/w/index.php?title=File:BSicon_KRZo.svg *License*: unknown *Contributors*: user:axpde

Image:Berlin SBahn HackescherMarkt east.jpg *Source*: http://en.wikipedia.org/w/index.php?title=File:Berlin_SBahn_HackescherMarkt_east.jpg *License*: unknown *Contributors*: User:H0tte

Image:berlin stadtbahn1.jpg *Source*: http://en.wikipedia.org/w/index.php?title=File:Berlin_stadtbahn1.jpg *License*: unknown *Contributors*: JuergenG

Image:T2-Savignyp.jpg *Source*: http://en.wikipedia.org/w/index.php?title=File:T2-Savignyp.jpg *License*: unknown *Contributors*: BLueFiSH.as, BlackIceNRW, Jcornelius, Kneiphof, Marcela, Platte, Urmelbeauftrager, 1 anonymous edits

Image:Berlin - S-Bahnhof Jannowitzbruecke.jpg *Source*: http://en.wikipedia.org/w/index.php?title=File:Berlin_-_S-Bahnhof_Jannowitzbruecke.jpg *License*: unknown *Contributors*: BLueFiSH.as, Jcornelius

Image:S-Bahn-Logo.svg *Source*: http://en.wikipedia.org/w/index.php?title=File:S-Bahn-Logo.svg *License*: unknown *Contributors*: User:LeCornichon

Image:Transitzug Bahnhof Zoo.jpg *Source*: http://en.wikipedia.org/w/index.php?title=File:Transitzug_Bahnhof_Zoo.jpg *License*: unknown *Contributors*: Mäfä

Image:S-Bahn Berlin Westkreuz Ringbahn.jpg *Source*: http://en.wikipedia.org/w/index.php?title=File:S-Bahn_Berlin_Westkreuz_Ringbahn.jpg *License*: unknown *Contributors*: Jcornelius

Image:Berlin B100 Kaiserdamm.jpg *Source*: http://en.wikipedia.org/w/index.php?title=File:Berlin_B100_Kaiserdamm.jpg *License*: unknown *Contributors*: User:Clemensfranz

File:Berliner Ringbahn1885.jpg *Source*: http://en.wikipedia.org/w/index.php?title=File:Berliner_Ringbahn1885.jpg *License*: unknown *Contributors*: Pyxlyst

File:Berlin-wall-map.png *Source*: http://en.wikipedia.org/w/index.php?title=File:Berlin-wall-map.png *License*: unknown *Contributors*: User:ChrisO, user:Grant

File:Berlin satellite image with Berlin wall.jpg *Source*: http://en.wikipedia.org/w/index.php?title=File:Berlin_satellite_image_with_Berlin_wall.jpg *License*: unknown *Contributors*: BLueFiSH.as, Beek100, Daniel 1992, Duesentrieb, Ed g2s, Emijrp, Krinkle, Magnus Manske, Paddy, Rüdiger Wölk, Sławojar, Srittau, Timeshifter, 4 anonymous edits

File:Bundesarchiv Bild 183-85458-0002, Berlin, Mauerbau, Kampfgruppen am Brandenburger Tor.jpg *Source*: http://en.wikipedia.org/w/index.php?title=File:Bundesarchiv_Bild_183-85458-0002,_Berlin,_Mauerbau,_Kampfgruppen_am_Brandenburger_Tor.jpg *License*: unknown *Contributors*: Junge, Peter Heinz

File:Berlin Wall 1961-11-20.jpg *Source*: http://en.wikipedia.org/w/index.php?title=File:Berlin_Wall_1961-11-20.jpg *License*: unknown *Contributors*: National Archives

File:Kennedy in Berlin.jpg *Source*: http://en.wikipedia.org/w/index.php?title=File:Kennedy_in_Berlin.jpg *License*: unknown *Contributors*: Robert Knudsen, White House, in the John F. Kennedy Presidential Library and Museum, Boston.

File:Berlin-Baby-Wall-1968.jpg *Source*: http://en.wikipedia.org/w/index.php?title=File:Berlin-Baby-Wall-1968.jpg *License*: unknown *Contributors*: User:Tamas Szabo

File:Karte berliner mauer en.jpg *Source*: http://en.wikipedia.org/w/index.php?title=File:Karte_berliner_mauer_en.jpg *License*: unknown *Contributors*: 32X, BLueFiSH.as, Beek100, Flying Saucer, Sansculotte, Sevela.p

File:Berlin - You are leaving.jpg *Source*: http://en.wikipedia.org/w/index.php?title=File:Berlin_-_You_are_leaving.jpg *License*: unknown *Contributors*: User:Staka

File:Berlin travel orders.png *Source*: http://en.wikipedia.org/w/index.php?title=File:Berlin_travel_orders.png *License*: unknown *Contributors*: W. B. Wilson

File:Conrad-schumann.jpg *Source*: http://en.wikipedia.org/w/index.php?title=File:Conrad-schumann.jpg *License*: unknown *Contributors*: Silversmith

File:Berlin-Memorial to the Victims of the Wall-1982.jpg *Source*: http://en.wikipedia.org/w/index.php?title=File:Berlin-Memorial_to_the_Victims_of_the_Wall-1982.jpg *License*: unknown *Contributors*: User:Lyricmac

File:ReaganBerlinWall.jpg *Source*: http://en.wikipedia.org/w/index.php?title=File:ReaganBerlinWall.jpg *License*: unknown *Contributors*: White House Photographic Office

File:BerlinWall-BrandenburgGate.jpg *Source*: http://en.wikipedia.org/w/index.php?title=File:BerlinWall-BrandenburgGate.jpg *License*: unknown *Contributors*: Sue Ream, photographer (San Francisco, California)

File:Juggling on the Berlin Wall 1a.jpg *Source*: http://en.wikipedia.org/w/index.php?title=File:Juggling_on_the_Berlin_Wall_1a.jpg *License*: unknown *Contributors*: User:Durova

File:Mauerspecht 1989.jpg *Source*: http://en.wikipedia.org/w/index.php?title=File:Mauerspecht_1989.jpg *License*: unknown *Contributors*: User:Skäpperöd

File:Bundesarchiv Bild 183-1989-1110-018, Berlin, Checkpoint Charlie, Nacht des Mauerfalls.jpg *Source*: http://en.wikipedia.org/w/index.php?title=File:Bundesarchiv_Bild_183-1989-1110-018,_Berlin,_Checkpoint_Charlie,_Nacht_des_Mauerfalls.jpg *License*: unknown *Contributors*: Oberst, Klaus

File:berlin fallofwall emerson.jpg *Source*: http://en.wikipedia.org/w/index.php?title=File:Berlin_fallofwall_emerson.jpg *License*: unknown *Contributors*: User:Maxonfire

File:Crane removed part of Wall Brandenburg Gate.jpg *Source*: http://en.wikipedia.org/w/index.php?title=File:Crane_removed_part_of_Wall_Brandenburg_Gate.jpg *License*: unknown *Contributors*: SSGT F. LEE CORKRAN

File:Bundesarchiv Bild 183-1990-0105-029, Berlin, Loch in Mauer am Reichstag.jpg *Source*: http://en.wikipedia.org/w/index.php?title=File:Bundesarchiv_Bild_183-1990-0105-029,_Berlin,_Loch_in_Mauer_am_Reichstag.jpg *License*: unknown *Contributors*: Reiche, Hartmut

File:Berlin wall 1990.jpg *Source*: http://en.wikipedia.org/w/index.php?title=File:Berlin_wall_1990.jpg *License*: unknown *Contributors*: user:Rnt20

File:Berlin wall at Potsdamer Platz March 2009.jpg *Source*: http://en.wikipedia.org/w/index.php?title=File:Berlin_wall_at_Potsdamer_Platz_March_2009.jpg *License*: unknown *Contributors*: User:Jlascar

File:Berlin_Wall_(13-8-2006).jpg *Source*: http://en.wikipedia.org/w/index.php?title=File:Berlin_Wall_(13-8-2006).jpg *License*: unknown *Contributors*: Beek100, Ddxc, DragonflySixtyseven, Erud, Herraotic, SatuSuro, 6 anonymous edits

File:Berlin wall 1.jpg *Source*: http://en.wikipedia.org/w/index.php?title=File:Berlin_wall_1.jpg *License*: unknown *Contributors*: User:Shreyasjoshis

File:Checkpoint Charlie Memorial.JPG *Source*: http://en.wikipedia.org/w/index.php?title=File:Checkpoint_Charlie_Memorial.JPG *License*: unknown *Contributors*: Original uploader was Victorgrigas at en.wikipedia

File:BERLINER MAUER 1961–1989 plaque.jpg *Source*: http://en.wikipedia.org/w/index.php?title=File:BERLINER_MAUER_1961–1989_plaque.jpg *License*: unknown *Contributors*: Kai81

Image:Wberlin transport 78.jpg *Source*: http://en.wikipedia.org/w/index.php?title=File:Wberlin_transport_78.jpg *License*: unknown *Contributors*: Original uploader was AlbertR at en.wikipedia

Image:Berlin Alert 0400.jpg *Source*: http://en.wikipedia.org/w/index.php?title=File:Berlin_Alert_0400.jpg *License*: unknown *Contributors*: Original uploader was Rw rynerson at en.wikipedia

File:JFK speech Ich bin ein berliner 1.jpg *Source*: http://en.wikipedia.org/w/index.php?title=File:JFK_speech_Ich_bin_ein_berliner_1.jpg *License*: unknown *Contributors*: Robert Knudsen, White House

File:Bundesarchiv B 145 Bild-F015843-0010, Berlin, Staatsbesuch Präsident der USA, Kennedy.jpg *Source*: http://en.wikipedia.org/w/index.php?title=File:Bundesarchiv_B_145_Bild-F015843-0010,_Berlin,_Staatsbesuch_Präsident_der_USA,_Kennedy.jpg *License*: unknown *Contributors*: Schmitt, Walter

File:Bundesarchiv Bild 183-L0331-0009, Potsdam, Grenzübergang Drewitz-Dreilinden, Verplomben von LKW.jpg *Source*: http://en.wikipedia.org/w/index.php?title=File:Bundesarchiv_Bild_183-L0331-0009,_Potsdam,_Grenzübergang_Drewitz-Dreilinden,_Verplomben_von_LKW.jpg *License*: unknown *Contributors*: Reiche, Hartmut

File:Bundesarchiv B 145 Bild-P004762, Berlin-Tempelhof, Flüchtlinge.jpg *Source*: http://en.wikipedia.org/w/index.php?title=File:Bundesarchiv_B_145_Bild-P004762,_Berlin-Tempelhof,_Flüchtlinge.jpg *License*: unknown *Contributors*: Arsenikk, Grayshi, KTo288, TommyG

Image:BerlinerBlockadeLuftwege.png *Source*: http://en.wikipedia.org/w/index.php?title=File:BerlinerBlockadeLuftwege.png *License*: unknown *Contributors*: Benutzer:LeerlaufprozessLeerlaufprozess

Image:Karte berliner mauer en.jpg *Source*: http://en.wikipedia.org/w/index.php?title=File:Karte_berliner_mauer_en.jpg *License*: unknown *Contributors*: 32X, BLueFiSH.as, Beek100, Flying Saucer, Sansculotte, Sevela.p

Image:Bundesarchiv Bild 173-1282, Berlin, Brandenburger Tor, Wasserwerfer.jpg *Source*: http://en.wikipedia.org/w/index.php?title=File:Bundesarchiv_Bild_173-1282,_Berlin,_Brandenburger_Tor,_Wasserwerfer.jpg *License*: unknown *Contributors*: Wolf, Helmut J.

File:Bundesarchiv B 145 Bild-P086547, Berlin, Sektorenübergang Oberbaumbrücke.jpg *Source*: http://en.wikipedia.org/w/index.php?title=File:Bundesarchiv_B_145_Bild-P086547,_Berlin,_Sektorenübergang_Oberbaumbrücke.jpg *License*: unknown *Contributors*: Origamiemensch

File:Bundesarchiv Bild 183-B1228-0010-003, Berlin, West-Berliner Besucher in Ostberlin.jpg *Source*: http://en.wikipedia.org/w/index.php?title=File:Bundesarchiv_Bild_183-B1228-0010-003,_Berlin,_West-Berliner_Besucher_in_Ostberlin.jpg *License*: unknown *Contributors*: Brüggmann, Eva

File:Bundesarchiv Bild 183-1989-1118-018, Berlin, Grenzübergang Bornholmer Straße.jpg *Source*: http://en.wikipedia.org/w/index.php?title=File:Bundesarchiv_Bild_183-1989-1118-018,_Berlin,_Grenzübergang_Bornholmer_Straße.jpg *License*: unknown *Contributors*: Roeske, Robert

File:Deutschland Bundeslaender 1957.png *Source*: http://en.wikipedia.org/w/index.php?title=File:Deutschland_Bundeslaender_1957.png *License*: unknown *Contributors*: David Kernow, Disposable.Heroes, Gauravjuvekar, Jed, Koavf, OwenBlacker, Trockennasenaffe, W-j-s, 1 anonymous edits

File:Volkspolizei at the official opening of the Brandenburg Gate.jpg *Source*: http://en.wikipedia.org/w/index.php?title=File:Volkspolizei_at_the_official_opening_of_the_Brandenburg_Gate.jpg *License*: unknown *Contributors*: SSGT F. Lee Corkran

File:Bundesarchiv B 145 Bild-F074398-0021, Bonn, Pressekonferenz Bundestagswahlkampf, Kohl.jpg *Source*: http://en.wikipedia.org/w/index.php?title=File:Bundesarchiv_B_145_Bild-F074398-0021,_Bonn,_Pressekonferenz_Bundestagswahlkampf,_Kohl.jpg *License*: unknown *Contributors*: Aibdescalzo, AnRo0002, Leit, Martin H., YMS, 1 anonymous edits

File:Cologne Reunification Kiss.png *Source*: http://en.wikipedia.org/w/index.php?title=File:Cologne_Reunification_Kiss.png *License*: unknown *Contributors*: Gary Mark Smith

File:UTafel Deutschland geteilt.jpg *Source*: http://en.wikipedia.org/w/index.php?title=File:UTafel_Deutschland_geteilt.jpg *License*: unknown *Contributors*: User:DorisAntony

File:Blank.png *Source*: http://en.wikipedia.org/w/index.php?title=File:Blank.png *License*: unknown *Contributors*: Bastique, Chlewey, ChrisDHDR, It Is Me Here, Jed, Paradoctor, Patrick, Penubag, Rocket000, Roomba, Timeroot, Tintazul

File:Flag of East Germany.svg *Source*: http://en.wikipedia.org/w/index.php?title=File:Flag_of_East_Germany.svg *License*: unknown *Contributors*: Flaggenentwurf: unbekannt diese Datei: Jwnabd

File:Coat of arms of East Germany.svg *Source*: http://en.wikipedia.org/w/index.php?title=File:Coat_of_arms_of_East_Germany.svg *License*: unknown *Contributors*: Wappenentwurf: Heinz Behling diese Datei: Jwnabd

File:East Germany 1956-1990.svg *Source*: http://en.wikipedia.org/w/index.php?title=File:East_Germany_1956-1990.svg *License*: unknown *Contributors*: User:StalwartUK, User:maix, user:Alphathon

File:Soviet Sector Germany.png *Source*: http://en.wikipedia.org/w/index.php?title=File:Soviet_Sector_Germany.png *License*: unknown *Contributors*: Alexandrin, Burts, Electionworld, Hanhil, Koavf, OwenBlacker, Razorbliss

File:Deutschland Bundeslaender 1949.png *Source*: http://en.wikipedia.org/w/index.php?title=File:Deutschland_Bundeslaender_1949.png *License*: unknown *Contributors*: Bukk, David Kernow, Gauravjuvekar, Hystrix, Jed, Koavf, Man vyi, OwenBlacker, Polletfa, W-j-s

File:Deutschland Besatzungszonen 1945.svg *Source*: http://en.wikipedia.org/w/index.php?title=File:Deutschland_Besatzungszonen_1945.svg *License*: unknown *Contributors*: user:glglgl

File:Bundesarchiv Bild 183-19000-3301, Berlin, DDR-Gründung, Wahl Pieck, Grotewohl.jpg *Source*: http://en.wikipedia.org/w/index.php?title=File:Bundesarchiv_Bild_183-19000-3301,_Berlin,_DDR-Gründung,_Wahl_Pieck,_Grotewohl.jpg *License*: unknown *Contributors*: Zühlsdorf

File:Fotothek df roe-neg 0002794 002 Portrait Walter Ulbrichts im Publikum der Bachfeier.jpg *Source*: http://en.wikipedia.org/w/index.php?title=File:Fotothek_df_roe-neg_0002794_002_Portrait_Walter_Ulbrichts_im_Publikum_der_Bachfeier.jpg *License*: unknown *Contributors*: Fredy.00, Groupsixty, High Contrast

File:Bundesarchiv Bild 183-R0518-182, Erich Honecker.jpg *Source*: http://en.wikipedia.org/w/index.php?title=File:Bundesarchiv_Bild_183-R0518-182,_Erich_Honecker.jpg *License*: unknown *Contributors*: A1B2C3D4, AnRo0002, Mbdortmund

Image:Karl Marx memorial.jpg *Source*: http://en.wikipedia.org/w/index.php?title=File:Karl_Marx_memorial.jpg *License*: unknown *Contributors*: Robert Steffens. Original uploader was RobbyBer at de.wikipedia

File:SED Logo.svg *Source*: http://en.wikipedia.org/w/index.php?title=File:SED_Logo.svg *License*: unknown *Contributors*: User:Connum, User:Stw

File:Bundesarchiv Bild 183-M0925-406, New York, Fahnen vor dem UNO-Gebäude.jpg *Source*: http://en.wikipedia.org/w/index.php?title=File:Bundesarchiv_Bild_183-M0925-406,_New_York,_Fahnen_vor_dem_UNO-Gebäude.jpg *License*: unknown *Contributors*: Spremberg, Joachim

File:Palast der Republik DDR 1977.jpg *Source*: http://en.wikipedia.org/w/index.php?title=File:Palast_der_Republik_DDR_1977.jpg *License*: unknown *Contributors*: Istvan

File:Nva-ehrenwache.jpg *Source*: http://en.wikipedia.org/w/index.php?title=File:Nva-ehrenwache.jpg *License*: unknown *Contributors*: Michail Jungierek

File:DDR Verwaltungsbezirke farbig.svg *Source*: http://en.wikipedia.org/w/index.php?title=File:DDR_Verwaltungsbezirke_farbig.svg *License*: unknown *Contributors*: Algos

File:Bundesarchiv Bild 183-1982-0816-002, Leipzig, Karl-Marx-Platz, Brunnen.jpg *Source*: http://en.wikipedia.org/w/index.php?title=File:Bundesarchiv_Bild_183-1982-0816-002,_Leipzig,_Karl-Marx-Platz,_Brunnen.jpg *License*: unknown *Contributors*: Gahlbeck, Friedrich

File:DDR economy-en.svg *Source*: http://en.wikipedia.org/w/index.php?title=File:DDR_economy-en.svg *License*: unknown *Contributors*: CIA (original), w:User:MysidMysid (SVG)

File:Bundesarchiv Bild 183-P0619-306, Trabant 601.jpg *Source*: http://en.wikipedia.org/w/index.php?title=File:Bundesarchiv_Bild_183-P0619-306,_Trabant_601.jpg *License*: unknown *Contributors*: Ludwig, Jürgen

File:Bertolt-Brecht.jpg *Source*: http://en.wikipedia.org/w/index.php?title=File:Bertolt-Brecht.jpg *License*: unknown *Contributors*: Kolbe, Jörg

File:Karin Janz.png *Source*: http://en.wikipedia.org/w/index.php?title=File:Karin_Janz.png *License*: unknown *Contributors*: DYNAMO

File:Stamps of Germany (DDR) 1974, MiNr 1951.jpg *Source*: http://en.wikipedia.org/w/index.php?title=File:Stamps_of_Germany_(DDR)_1974,_MiNr_1951.jpg *License*: unknown *Contributors*: User:Nightflyer

File:1989 CPA 6119.jpg *Source*: http://en.wikipedia.org/w/index.php?title=File:1989_CPA_6119.jpg *License*: unknown *Contributors*: User:Mariluna

Image:Soviet War Memorial.JPG *Source*: http://en.wikipedia.org/w/index.php?title=File:Soviet_War_Memorial.JPG *License*: unknown *Contributors*: Treptower, Wst

Image:Treptower Park Pedestal.JPG *Source*: http://en.wikipedia.org/w/index.php?title=File:Treptower_Park_Pedestal.JPG *License*: unknown *Contributors*: Victorgrigas

file:Berlin - Bahnhof Warschauer Straße, nach Abriss.jpg *Source*: http://en.wikipedia.org/w/index.php?title=File:Berlin_-_Bahnhof_Warschauer_Straße,_nach_Abriss.jpg *License*: unknown *Contributors*: BLueFiSH.as, Jcornelius, Rabenkind

Image:Berlin - U-Bahnhof Warschauer Straße.jpg *Source*: http://en.wikipedia.org/w/index.php?title=File:Berlin_-_U-Bahnhof_Warschauer_Straße.jpg *License*: unknown *Contributors*: BLueFiSH.as, Jcornelius, Rabenkind

Image:S-Bahn Warschauerstrasse west.jpg *Source*: http://en.wikipedia.org/w/index.php?title=File:S-Bahn_Warschauerstrasse_west.jpg *License*: unknown *Contributors*: Angel Perea Martinez (angelpeream@yahoo.com)

Image:Warschauer strasse strassenschild.jpg *Source*: http://en.wikipedia.org/w/index.php?title=File:Warschauer_strasse_strassenschild.jpg *License*: unknown *Contributors*: Jcornelius, Klugschnacker, MB-one, Mattes

MIX
Papier aus verantwortungsvollen Quellen
Paper from responsible sources
FSC® C105338

Printed by Books on Demand GmbH, Norderstedt / Germany